VALVE AND TRANSISTOR
AUDIO AMPLIFIERS

VALVE AND TRANSISTOR AUDIO AMPLIFIERS

JOHN LINSLEY HOOD

Newnes

AMSTERDAM BOSTON HEIDELBERG LONDON NEW YORK OXFORD
PARIS SAN DIEGO SAN FRANCISCO SINGAPORE SYDNEY TOKYO

Newnes
An imprint of Elsevier
Linacre House, Jordan Hill, Oxford OX2 8DP
30 Corporate Drive, Burlington, MA 01803

First published 1997
Reprinted 2000, 2001, 2003, 2006

British Library Cataloguing in Publication Data
A catalogue record for this book is available from the British Library

Library of Congress Cataloguing in Publication Data
A catalogue record for this book is available from the Library of Congress

ISBN 0 7506 3356 5

For information on all Newnes publications
visit our website at www.newnespress.com

Working together to grow
libraries in developing countries
www.elsevier.com | www.bookaid.org | www.sabre.org

ELSEVIER   BOOK AID   Sabre Foundation
            International

Typeset by Design 2 Print, Droitwich, Worcestershire

CONTENTS

PREFACE

An old friend of mine was in the habit of remarking that an engineer is someone who can do for five pounds what any damn fool can do for fifty. Needless to say, he was an engineer. I suppose, in a sense, that this is a fairly good description of the major aims of engineering practice; that the designer should seek elegant, efficient and cost-effective ways of achieving a clearly specified objective. Only when satisfactory ways of doing this cannot be found is there a need to seek more elaborate or costly ways to get this result.

In the field of audio amplifiers there has been a great interest in techniques for making small electrical voltages larger ever since mankind first attempted to transmit the human voice along lengthy telephone cables. This quest received an enormous boost with the introduction of radio broadcasts, and the resulting mass-production of domestic radio receivers intended to operate a loudspeaker output. However, the final result, in the ear of the listener, though continually improved over the passage of the years, is still a relatively imperfect imitation of the real-life sounds which the engineer has attempted to copy. Although most of the shortcomings in this attempt at sonic imitation are not due to the electronic circuitry and the amplifiers which have been used, there are still some differences between them, and there is still some room for improvement.

In this book I have looked at the audio amplifier designs which have been developed over the past fifty years, in the hope that the information may be of interest to the user or would-be designer, and I have tried to explore both the residual problems in this field, and the ways by which these may be lessened.

I believe, very strongly, that the only way by which improvements in these things can be obtained is by making, and analysing, and recording for future use, the results of instrumental tests of as many relevant aspects of the amplifier electrical performance as can be devised. Obviously, one must not forget that the final result will be judged in the ear of the listener, so that, when all the purely instrumental tests have been completed, and the results judged to be satisfactory, the equipment should also be assessed for sound quality, and the opinions in this context of as many interested parties as possible should be canvassed.

Listening trials are difficult to set up, and hard to purge of any inadvertent bias in the way equipment is chosen or the tests are carried out. Human beings are also notoriously prone to believe that their preconceived views will prove to be correct. The tests must therefore be carried out on a double blind basis, when neither the listening panel, nor the persons selecting one or other of the items under test, know what piece of hardware is being tested.

If there is judged to be any significant difference in the perceived sound quality, as between different pieces of hardware which are apparently identical in their measured performance, the type and scope of the electrical tests which have been made must be considered carefully to see if any likely performance factor has been left unmeasured, or not given adequate weight in the balance of residual imperfections which exist in all real-life designs.

A further complicating factor arises because some people have been shown to be surprisingly sensitive to apparently insignificant differences in performance, or to the presence of apparently trifling electrical defects – not always the same ones – so, since there are bound to be some residual defects in the performance of any piece of hardware, each listener is likely to have his or her own opinion of which of these sounds best, or which gives the most accurate reproduction of the original sound – if this comparison is possible.

The most that the engineer can do, in this respect, is to try to discover where these performance differences arise, or to help decide the best ways of getting the most generally acceptable performance.

It is simple to specify the electrical performance which should be sought. This is that, for a signal waveform which does not contain any frequency components which fall outside the audio frequency spectrum – which may be defined as 10Hz–20kHz – there should be no measurable differences, except in amplitude, between the waveform present at the input to the amplifier or other circuit layout (which must be identical to the waveform from the signal source before the amplifier or other circuit is connected to it) and that present across the circuit output to the load when the load is connected to it.

In order to achieve this objective, the following requirements must be met.

- The constant amplitude (±0.5dB) bandwidth of the circuit, under load, and at all required gain and output amplitude levels, should be at least 20Hz–20kHz.

- The gain and signal to noise ratio of the circuit must be adequate to provide an output signal of adequate amplitude, and that the noise or other non-signal related components must be inaudible under all conditions of use.

- Both the harmonic and intermodulation distortion components present in the output waveform, when the input signal consists of one or more pure sinusoidal waveforms within the audio frequency spectrum, should not exceed some agreed level. (In practice, this is very difficult to define because the tolerable magnitudes of such waveform distortion components depend on their frequency, and also, in the case of harmonic distortion, on the order (i.e. whether they are 2nd, 3rd, 4th or 5th as the case may be). Contemporary thinking is that all such distortion components should not exceed 0.02%, though, in the particular case of the 2nd harmonic, it is probably undetectable below 0.05%.)

- The phase linearity and electrical stability of the circuit, with any likely reactive load, should be adequate to ensure that there is no significant alteration of the form of a transient or discontinuous waveform such as a fast square or rectangular wave, provided that this would not constitute an output or input overload. There should be no ringing (superimposed spurious oscillation) and, ideally, there should also be no waveform overshoot, under square-wave testing, in which the signal should recover to the undistorted voltage level, ±0.5%, within a settling time of 20μs.

- The output power delivered by the circuit into a typical load – bearing in mind that this may be either higher or lower than the nominal impedance at certain parts of the audio spectrum – must be adequate for the purpose required.

- If the circuit is driven into overload conditions, it must remain stable, the clipped waveform should be clean and free from instability, and should recover to the normal signal waveform level with the least possible delay – certainly less than 20μs.

In addition to these purely electrical specifications, which would probably be difficult to meet, even in a very high quality solid state design – and most unlikely to be satisfied in any transformer coupled system – there are a number of purely practical considerations, such as that the equipment should be efficient in its use of electrical power, that its heat dissipation should not present problems in housing the equipment, and that the design should be cost-effective, compact and reliable.

Since it is improbable that all these performance requirements will be met, in any practical design, it is implicit that the designer will have made certain performance compromises, in which better performance in certain respects has been traded off against a lesser degree of excellence in others. For myself, I think that the total harmonic or intermodulation distortion, which is frequently specified in the makers data sheets, is less important in determining the tonal quality of the system, provided that it is better than 0.05%, than its behaviour under transient conditions, when tested with typical (or accurately simulated) reactive loads, performance in which is seldom or never quoted. Where appropriate, in the following text, I will try to show where various benefits are obtained at the cost of some other potential drawbacks.

JLH 1996.

CHAPTER 1

ACTIVE COMPONENTS

Electronic amplifiers are built up from combinations of active and passive components. The active ones are those, like valves or transistors or integrated circuits, that draw electrical current from suitable voltage supply lines and then use it to generate or modify some electrical signal. The passive components are those, like capacitors, resistors, inductors, potentiometers or switches, which introduce no additional energy into the circuit, but which act upon the input or output voltages and currents of the active devices in order to control the way they operate. Of these, the active components are much more fun, so I will start with these.

Although the bulk of modern electronic circuitry is based on 'solid state' components, for very good engineering reasons – one could not, for example, build a compact disc player using valves, and still have room in one's house to sit down and listen to it – all the early audio amplifiers were based on valves, and it is useful to know how these worked, and what the design problems and circuit options were, in order to get a better understanding of the technology. Also, there is still an interest on the part of some 'Hi-Fi' enthusiasts in the construction and use of valve operated audio amplifiers, and additional information on valve based circuitry may be welcomed by them.

Valves or Vacuum Tubes

The term thermionic valve (or valve for short) was given, by its inventor, Sir Ambrose Fleming, to the earliest of these devices, a rectifying diode. Fleming chose the name because of the similarity of its action, in allowing only a one-way flow of current, to that of a one-way air valve on an inflatable tyre, and the way it operated was by controlling the internal flow of thermally generated electrons, which he called 'thermions', hence the term thermionic valve. In the USA they are called 'vacuum tubes'. These devices consist of a heated cathode, mounted, in vacuum, inside a sealed glass or metal tube. Other electrodes, such as anodes or grids are then arranged around the cathode, so that various different functions can be performed.

The descriptive names given to the various types of valve are based on the number of its internal electrodes, so that a valve with two electrodes (a cathode and an anode) will be called a 'diode', one with three electrodes (a cathode, a grid and an anode) will be called a 'triode', one with four (a cathode, two grids and an anode) will be called a 'tetrode', and so on.

It helps to understand the way in which valves work, and how to get the best performance from them, if one understands the functions of these internal electrodes, and the way in which different groupings of them affect the characteristics of the valve, so, to this end, I have listed them, and examined their functions separately.

The cathode

This component is at the heart of any valve, and is the source of the electrons with which it operates. It is made in one or other of two forms: either a short length of resistor wire, made of nickel, folded into a 'V' shape, and supported between a pair of stiff wires at its base and a light tension spring at its top, as shown in Figure 1.1a, or a metallic tube, usually made of nickel, with a bundle of nickel or tungsten heater wires gathered inside it, as shown in Figure 1.1b. Whether the cathode is a directly heated 'filament' or an indirectly heated metal cylinder, its function and method of operation is the same, though, other things being equal, the directly heated filament is much more efficient, in terms of the available electron emission from the cathode in relation to the amount of power required to heat it to its required operating temperature (about 775°C for one having an oxide coated construction).

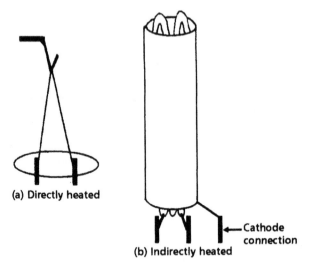

(a) Directly heated

Cathode connection

(b) Indirectly heated

Figure 1.1
Valve cathode styles

It is possible to use a plain tungsten filament as a cathode, but it needs to be heated to some 2500°C to be usable, and this requires quite a substantial amount of power, and leads to other problems such as fragility. Virtually all contemporary low to medium power valves use oxide coated cathodes, which are made from a mixture of the oxides of calcium, barium, and strontium deposited on a nickel substrate.

In the manufacture of the valve these chemicals are applied to the cathode as a paste composed of a binding agent, the metals in the form of their carbonates, and some small quantities of doping agents, typically of rare-earth origin. The metal carbonates are then reduced to their oxides by subsequent heating during the last stage in the process of evacuating the air from the valve envelope.

In use, a chemical reaction occurs between the oxide coating and the heated nickel cathode tube (or the directly heated filament) which causes the alkali metal oxides to be locally reduced to the free metal, which then slowly diffuses out to the cathode surface to form the electron emitting layer. The extent of electronic emission from the cathode depends critically upon its temperature, and the value chosen for this in practice is a compromise between performance and life expectancy, since higher cathode temperatures lead to shorter cathode life, due to the loss through evaporation of the active cathode metals, while a lower limit to the working temperature is set by the need to have an adequate level of electron emission.

When hot, the cathode will emit electrons, which form a cloud around it, a situation in which the thermal agitation of the electrons in the cathode body, which causes electrons to escape from its surface, is balanced by the growing positive charge which the cathode has acquired as the result of the loss of these electrons. This electron cloud is called the 'space charge', and it plays an important part in the operation of the valve; a matter which is discussed later.

The anode

In the simplest form of valve, the diode, the cathode is surrounded by a metal tube or box, called the anode or plate. This is usually made of nickel, and it will attract electrons from the space charge if it is made positive with regard to the cathode. The amount of current which will flow depends on the closeness of the anode box to the cathode, the effective area of the cathode, the voltage on the anode, and the cathode temperature. For a fixed cathode temperature and anode voltage the ratio of anode voltage to current flow determines the anode current resistance, R_a, which is measured by the current flow for a given applied voltage – as shown in the equation

$$R_a = dV_a/dI_a$$

Because the anode is bombarded by electrons accelerated towards it by the applied anode voltage, when they collide with the anode their kinetic energy is converted into heat, which raises the anode temperature. This heat evolution is normally unimportant, except in the case of power rectifiers or power output valves, when care should be taken to ensure that the makers' current and voltage ratings are not exceeded. In particular, there is an inherent problem that if the anode becomes too hot, any gases which have been trapped in pores within its structure will be released, and this will impair the vacuum within the valve, which can lead to other problems.

The control grid

If the cathode is surrounded by a wire grid or mesh – in practice, this will usually take the form of a spiral coil, spot-welded between two stiff supporting wires, of the form shown in Figure 1.2 – the current flow from the cathode to the anode can be controlled by the voltage applied to the grid, such that if the grid is made positive, more negatively charged electrons will be attracted away from the cathode and encouraged to continue on their way to the anode. On the other hand, if the grid is made negative, it will repel the electrons emitted by the cathode, and reduce the current flow to the anode.

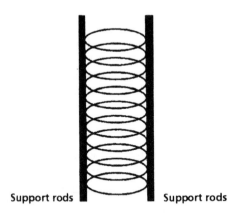

Support rods   Support rods

Figure 1.2
Control grid construction

It is this quality which is the most useful aspect of a valve, in that a quite large anode current flow can be controlled by a relatively small voltage applied to the grid, and so long as the grid is not allowed to swing positive with respect to the cathode, no current will flow in the grid circuit, and its effective input impedance at low frequencies will be almost infinite. This ability to regulate a large current at a high voltage by a much smaller control voltage allows the valve to amplify small electrical signals, and since the relationship between grid voltage and anode current is relatively linear, as shown in Figure 1.3, this amplification will cause relatively little distortion in the amplified signal. The theoretical amplification factor of a valve, operating into an infinitely high impedance anode load, is denoted by the Greek symbol μ.

Although in more complex valves there may be several grids between the cathode and the anode, the grid which is closest to the cathode will have the greatest influence on the anode current flow, and this is therefore usually called the control grid.

The effectiveness of the grid in regulating the anode current depends on the relative proximity of the grid and the anode to the cathode, in that, if the grid is close to the cathode, but the anode is relatively remote, the effectiveness of the grid in determining the anode current will be much greater, and will therefore give a higher value of μ than if the anode is closer to the grid and cathode. Unfortunately, there is a snag in that the anode current resistance of the valve, R_a, is also related to the anode/cathode spacing, and becomes higher as the anode/cathode spacing is increased. The closeness of the pitch of the wire spiral which forms the grid also affects the anode current resistance in that a close spacing will lead to a high R_a, and vice versa.

The stage gain (M) of a simple valve amplifier, of the kind shown in Figure 1.4, is given by the equation

$$M = \mu R/(R + R_a)$$

so that a low impedance valve, such as a 6SN7 (typical $I_a = 9mA$, $R_a = 7.7k$, $\mu = 20$), which has close anode–grid and grid–cathode spacings, and a relatively open pitch in the grid wire spiral, will have a high possible anode current but a low amplification factor, while a high impedance valve such as a 6SL7 (typical $I_a = 2.3mA$, $R_a = 44k$, $\mu = 70$) will have a low stage gain unless the circuit used has a high value of anode load resistance (R), and this, in turn, will demand a high value of HT voltage.

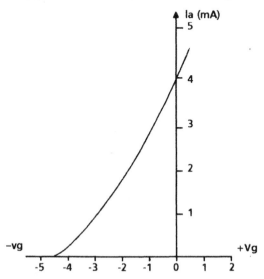

Figure 1.3
Triode valve characteristics

The space charge

Although a cloud of electrons will surround any heated cathode mounted in a vacuum, and will act as a reservoir of electrons when these are drawn off as anode current, their

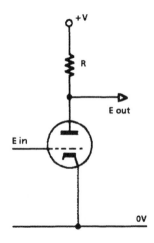

Figure 1.4
Simple valve amplifier

presence becomes of particular importance when a negatively charged control grid is introduced into the system, in that the electron cloud will effectively fill the space between the cathode and the grid, and will act as the principal source of electrons.

The presence of this electron cloud – known as the space charge – has several important operational advantages. Of these the first is that, by acting as an electron reservoir, it allows larger, brief duration, current flows than would be available from the cathode on its own, and that it acts as a measure of protection to the cathode against the impacts of positive ions created by electronic collisions with the residual gases in the envelope, since these ions will be attracted towards the more negatively charged cathode. Finally, left to itself, the electronic emission from the cathode suffers from both 'shot' and 'flicker' noise, a current fluctuation which is averaged out if the anode current is drawn from the space charge.

This random emission of electrons from a space charge depleted cathode is used to advantage in a 'noise diode', a wide-band noise source which consists of a valve in which the cathode is deliberately operated at a low temperature to prevent a space charge from forming, so that a resultant noisy current can be drawn off by the anode.

In the case of a triode used as an output power valve, where large anode currents are needed, the grid mesh must be coarse, and the grid–cathode spacing must be close. This limits the formation of an adequate space charge in the grid–cathode gap, and, in its absence, the cathode must have a higher emission efficiency than would be practicable with an indirectly heated system, and this means that a directly heated filament must be used instead. Usually, the filament voltage will be low to minimise cathode induced 'hum', and the filament current will be high, because of the size of the filament (2.5A at 2.5V in the case of the 2A3 valve).

Directly heated cathodes are also commonly used in valve HT rectifiers, such as the 5U4 or the 5Y3, because the higher cathode emission reduces the voltage drop across the valve and increases the available HT output voltage by comparison with a similar power supply using an indirectly heated cathode type.

Tetrodes and pentodes

Although the triode valve has a number of advantages as an amplifier – such as a low noise and low distortion factor – it suffers from the snag that there will be a significant capacitance, typically of the order of 2.5pF, between the grid and the anode. In itself, this latter capacitance would seem to be too small to be troublesome, but, in an amplifying stage with a gain of, say, 100, the Miller effect will increase the capacitance by a factor of 101, increasing the effective input capacitance to 252.5pF, which could influence the performance of the stage.

When triode valves were used as RF amplifiers, in the early years of radio, this anode–grid capacitance caused unwanted RF instability, and the solution which was adopted was the introduction of a 'screening' grid between the triode control grid and its anode, which reduced this anode–grid capacitance, in the case of a screened grid or tetrode valve, to some 0.025pF.

A further effect that the inclusion of a screening grid had upon the valve characteristics was to make the anode current, in its linear region, almost independent of the anode voltage, which led to very high values for R_a and μ. Unfortunately, the presence of this grid caused a problem that when the anode voltage fell, during dynamic conditions, to less than that of the screening grid, electrons hitting the anode could cause secondary electrons to be ejected from its surface – especially if the anode was hot or its surface had been contaminated by cathode material – and these would be collected by the screening grid, which would cause a kink in the anode current/voltage characteristics. While this might not matter much in an RF amplifier, it would cause an unacceptable level of distortion if used in an audio amplifier stage.

Two solutions were found for this problem, of which the simplest was to interpose an additional, open mesh, grid between the anode and the screening grid. This grid will normally be connected to the cathode, either externally or within the valve envelope, and is called the suppressor grid because it acts to suppress the emission of secondary electrons from the anode.

Since this type of valve had five electrodes it was called a 'pentode'. A typical small-signal pentode designed specifically for use in audio systems is the EF86, in which steps have also been taken to reduce the problem of microphony when the valve is used in the early stages of an amplifying system. The EF86 also has a wire mesh screen inside the glass envelope, and surrounding the whole of the electrode structure. This is connected to pins 2 and 7, and is intended to lessen the influence of external voltage fields on the electron flow between the valve electrodes.

In use, a small-signal pentode amplifying stage will give a much higher stage gain than a medium impedance triode valve (250˜ in comparison with, say, 30˜). It will also have a better HF gain due to its lower effective anode–grid capacitance. However, a triode gain stage will probably have a distortion figure, other things being equal, which is about half that of a pentode.

The second solution to the problem of anode current non-linearity in tetrodes, particularly suited to the output stages of audio amplifiers, was the alignment of the wires of the control grid and screening grid so that they constrained the electron flow into a series of beams, which served to sweep any secondary electrons back towards the anode – a process which was helped by the inclusion within the anode box of a pair of 'beam confining electrodes', which modified the internal electrostatic field pattern. These are internally connected to the cathode, and take the form shown in Figure 1.5. These valves were called beam-tetrodes or kinkless tetrodes, and had a lower distortion than output pentodes. Valves of this type, such as the 6L6, the 807, the KT66 and KT88, were widely employed in the output stages of the high quality audio amplifiers of the 1950s and early 1960s.

Both pentodes and beam-tetrodes can be used with their screen grids connected to their anodes. In this mode their characteristics will resemble a triode having a similar grid–cathode and grid–anode spacings to the grid–cathode and grid–screen grid spacings of the pentode. The most common use of this form of connection is in power output stages, where a triode connected beam tetrode will behave much like a power triode, without the need for a directly heated (and hum-inducing) cathode.

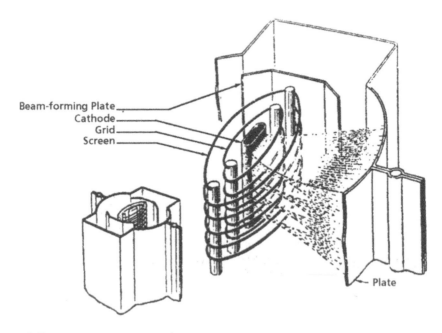

Beam-forming Plate
Cathode
Grid
Screen

Plate

Figure 1.5
Construction of beam-tetrode (by courtesy of RCA)

Valve parameters

In addition to the anode current resistance, R_a, and the amplification factor, μ, mentioned above, there is also the valve slope or mutual conductance (g_m) which is a measure of the extent to which the anode current will be changed by a change in grid voltage. Traditionally this would be quoted in milliamperes per volt (mA/V or milli-Siemens, written as mS), and would be a useful indication of the likely stage gain which would be given by the valve in an amplifying circuit.

This would be particularly helpful in the case of a pentode amplifying stage, where the value of R_a would probably be very high in comparison with the likely value of load resistance. (For example, in the case of the EF86, R_a is quoted as 2.5MΩ and the g_m is 2mA/V.) In this case, the stage gain (M) can be determined, approximately, by the relationship $M \approx -g_m \cdot R_L$, which, for a 100k anode load would be $\approx -200^-$.

The various valve characteristics are defined mathematically as,

$R_a = dV_a/dI_a$ at a constant grid voltage,
$g_m = -dI_a/dV_g$ at a constant anode voltage, and
$\mu = -dV_a/dV_g$ at a constant anode current.

In these equations the negative sign takes account of the phase inversion of the signal. These parameters are related to one another by the further equation,

$$g_m = \mu/R_a \text{ or } \mu = g_m R_a$$

Gettering

The preservation of a high vacuum within its envelope is essential to the life expectancy and proper operation of the valve. However, it is difficult to remove all traces of residual gas on the initial pumping out of the envelope, quite apart from the small but continuing gas evolution from the cathode, or any other electrodes which may become hot in use. The solution to this problem is the inclusion of a small container, known as a boat, mounted somewhere within the envelope, but facing away from the valve electrodes, which contains a small quantity of reactive material, such as metallic calcium and magnesium.

The boat is positioned so that after the pumping out of the envelope has been completed, and the valve had been sealed off, the getter could be caused to evaporate on to the inner face of the envelope by heating the boat with an induction heating coil. Care is taken to ensure that as little as possible of the getter material finds its way on to the inner faces of the valve electrodes, where it may cause secondary emission, or on to the mica spacers where it may cause leakage currents between the electrodes.

While this technique is reasonably effective in cleaning up the gas traces which arise during the use of the valve, the vacuum is never absolute, and evidence of the residual gas can sometimes be seen as a faint, deep blue glow in the space within the anode envelope of a power output valve. If, however, there is a crack in the glass envelope, or some other cause of significant air leakage into the valve interior, this will become apparent because of a whitening of the edges of the normally dark, mirror-like surface of the getter deposit on the inside of the valve envelope. A further sign of the ingress of air into the valve envelope is the presence of a pinkish-violet glow which extends beyond the confines of the anode box. By this time the valve must be removed and discarded, to prevent damage to other circuit components through an increasing and uncontrolled current flow.

Cathode and heater ratings

For optimum performance, the cathode temperature should be maintained, when in use, at its optimum value, and this requires that the heater or filament voltages should be set at the correct levels. Since the voltage of the domestic AC power supply is not constant, the design ratings for the heater or filament supply must take account of this. However, this is not as difficult to do as it might appear. For example, Brimar, a well-known valve manufacturer, make the following recommendations in their *Valve and Teletube Manual*:'the heater supply voltages should be within ±5% of the rated value when the heater transformer is fed with its nominal input voltage, provided that the mains power supply is within ±10% of its declared value'.

An additional requirement is that, because of inevitable cathode-heater leakage currents, the voltages between these electrodes should be kept as low as possible, and should not exceed 200V. Moreover, there must always be a resistive path, not exceeding 250kΩ, between the cathode and heater circuits. (As can be seen from some of the audio amplifier circuits shown in Chapter 6, electronic circuit designers often neglected this particular piece of advice.)

As a practical point, the wiring of the heater circuit, which is usually operated at 6.3V AC, will normally be installed as a twisted pair to minimise the induction of mains hum into sensitive parts of the system, as will the heater wiring inside the cathode tube of low noise valves, such as the EF86. With modern components, such as silicon diodes and low-cost regulator ICs, there is no good reason why the heater supplies to high quality valve amplifiers should not be derived from smoothed and stabilised DC sources.

It has been suggested that the cathodes of valves can be damaged by reverse direction ionic bombardment if the HT voltage is applied before the cathode has had a chance to warm up and form a space charge, and that the valve heaters should be left on to avoid this problem. In practice, this problem does not arise because gaseous ions are only formed by collisions between residual gas molecules and the electrons in the anode current stream. If the cathode has not reached operating temperature there will be little or no anode current, and, consequently, no gaseous ions produced as a result of it. Brimar specifically warn against leaving the cathode heated, in the absence of anode current, in that this may lead to cathode poisoning, because of chemical reactions occurring between the exposed reactive metal of the cathode surface and any gaseous contaminants present within the envelope. Unfortunately, the loss of electron emissivity as the cathode temperature is reduced occurs more rapidly than the reduction in the chemical reactivity of the cathode metals.

Indirectly heated HT rectifier valves have been used, in spite of their lower operating efficiency, to ensure that the full HT voltage was not applied to the equipment before the other valves had warmed up, but this was done to avoid the HT rail over-voltage surge which would otherwise occur, and allow the safe use of lower working voltage, and less expensive, components such as HT reservoir, smoothing, or inter-valve coupling capacitors.

Microphony

Any physical vibration of the grid (or filament, in the case of a directly heated cathode) will, by altering the grid–cathode spacing, cause a fluctuation of the anode current, and this will cause an audible ringing sound when the envelope is tapped – an effect known as microphony in the case of a valve used in audio circuitry. Great care must therefore be taken in the manufacture of valves to maintain the firmness of the mounting of the grids and other electrodes. This is done by the use of rigid supporting struts whose ends are located in holes punched in stiff mica disc shaped spacers which, in turn, are a tight fit within the valve envelope.

Since a microphonic valve will pick up vibration from any sound source, such as a loudspeaker system in proximity to it, and convert these sounds into (inevitably distorted) electrical signals which will be added to the amplifier output, this can be a significant, but unsuspected, source of signal distortion which will not be revealed during laboratory testing on a resistive dummy load. Since it is difficult to avoid valve microphony completely, and it is equally difficult to sound proof amplifiers, this type of distortion will always occur unless such valve amplifier systems are operated at a low volume level or the amplifier is located in a room remote from the loudspeakers.

Solid State Devices

Bipolar junction transistors

'N'- and 'P'- type materials

Most materials can be grouped in one or other of three classes, insulators, semiconductors or conductors, depending on the ease or difficulty with which electrons can pass through them. In insulators, all of the electrons associated with the atomic structure will be firmly bound in the valency bands of the material, while in good, usually metallic, conductors many of the atomic electrons will only be loosely bound, and will be free to move within the body of the material.

In semiconductors, at temperatures above absolute zero ($0°K$ or $-273.15°C$), electrons will exist in both the valency levels where they are not free to leave the atoms with which they are associated, and in the conduction band, in which they are free to travel within the body of the material. This characteristic is greatly influenced by the 'doping' of the material, which is normally done, during the manufacture of the semiconductor material, by introducing carefully controlled amounts of specific impurities into the molten mass from which the single semiconductor crystal is grown. The most common semiconductor material in normal use is silicon, because it is cheap and readily available, and has good thermal properties. Germanium, the material from which all early transistors were made, has electrical characteristics which are greatly influenced by its temperature, which is inconvenient in use. Also it does not lend itself at all well to contemporary mass-production techniques.

In the case of silicon, which has very little conductivity in its undoped 'intrinsic' form, the most common dopants are boron or aluminium which give rise to a semiconductor with a deficiency of valency electrons, usually referred to as holes – called a 'P'- type material – or phosphorus, which will cause the silicon to have a surplus of valency electrons, which forces some of them into the conduction band. Such a semiconductor material would be termed 'N' type. Both P- type and N- type silicon can be quite highly conductive, depending on the doping levels used.

Fermi levels

The electron energy distribution in single-crystal P- and N- type materials is shown in Figure 1.6, and the mean electron energy levels, known as the Fermi levels, are shown

as dotted lines on the diagrams. If two such differently doped semiconductor materials
are in intimate contact with one another, electrons will diffuse across the junction so
that the Fermi level is the same on both sides. This will cause a shift in the relative
electrical potential of the two doped regions as is shown in Figure 1.7, and a number
of interesting effects arise from this.

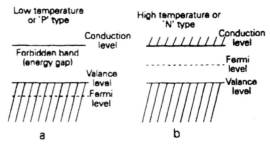

Figure 1.6
Average electron energy levels in semi-conductors as a function of temperature or
doping level

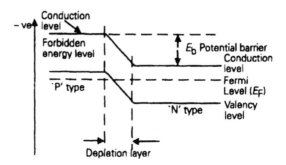

Figure 1.7
Mechanism by which potential barrier arises at P-N junction

(In practice, the only way by which such an intimate contact could be achieved within
a single crystal of a semiconductor material is by diffusing, say, a P-type impurity into
a wafer of N-type material, so that where these doped regions came into contact there
would be a predominance of P-type silicon on one side of the boundary and of N-type
silicon on the other.)

The P-N potential barrier and the depletion zone

The first of the phenomena which occur at the P-N junction is that the electron
conduction bands are displaced, so that electrical current (which consists, in the
philosophy of this book, of a movement of electrons from one place to another) cannot
flow through a semiconductor junction, even in the forward direction, from the
electron rich N-type zone into the electron deficient P-type one, until a high enough
potential exists to overcome the potential barrier (the voltage gap between the two

conduction bands). This is about 0.58V in a silicon junction at room temperature, and decreases as the junction temperature is increased.

The second effect, shown graphically in Figure 1.8, is the creation of a depletion zone, in which the diffusion of electrons across the junction from the electron rich N region into the electron deficient P region fills in the notional holes (labelled with '+' signs in the drawing) where electrons should have been but were not. This leaves a region where there are no electrons at all, and current will not flow through it. The effect on the width of the depletion zone of the voltage across the junction is shown, schematically, in Figure 1.9.

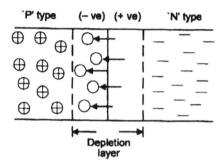

Figure 1.8
Growth of depletion layer on either side of a P-N junction

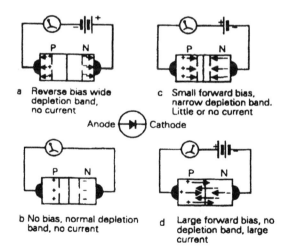

Figure 1.9
Influence of external potential on width of depletion zone in junction 'diode'

The semiconductor diode

The effect of the voltage across a semiconductor junction, in this case silicon, on its current flow is illustrated in Figure 1.10. When a forward (i.e. conduction direction) voltage is applied across such a P-N junction, little current will flow until the voltage exceeds the potential barrier, when the current flow will increase rapidly until any further increase is limited by the conducting resistance of the semiconductor material or of the external circuit.

In the reverse or non-conducting direction there will always be a small leakage current which will gradually increase as the reverse potential is increased, up to the voltage at which the junction breaks down. This breakdown is due to what is termed an avalanche effect, in which the electrons which form the reverse leakage current, accelerated by the reverse potential, reach a high enough kinetic energy to cause ionisation in the junction material. This ionisation releases further electrons which, in turn, will be involved in further collisions, increasing yet further the leakage current.

In general, such reverse breakdown effects are to be avoided because the heat involved causes a temperature rise at the junction which can permanently damage the diode. However, some diodes are deliberately manufactured, by the use of very high doping levels, so that reverse breakdown occurs at voltages which are low enough for thermal damage to be avoided. These are commonly called Zener diodes, though this term is really only appropriate for diodes having a reverse breakdown voltage below about 5.5V.

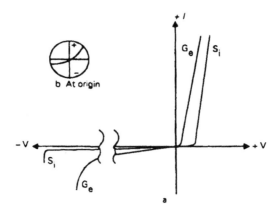

Figure 1.10
Normal forward and reverse characteristics of silicon and germanium junction diodes

Junction capacitance

An additional feature of the semiconductor junction is its capacitance. If the junction is reverse biased this capacitance is relatively small, and has the characteristics shown, approximately, by the relationship

$$C \approx k/\sqrt{V}$$

where V is the voltage across the junction, and k is some constant which depends on the doping level of the material, and the area and temperature of the junction. The capacitance of a forward biased junction is much larger, and much more strongly dependent on the doping level, the junction temperature and the forward current.

Typical values of capacitance for a reverse biased small-signal diode will be of the order of 2–15pF, while in the case of a forward biased junction the capacitance could well be some ten times this value, depending on the diode current. (In the case of transistors, where the relationship is much more complicated, the manufacturers seldom attempt to specify the input (base-emitter) capacitance, but lump this factor, along with other things which affect the transistor HF performance, in a general term, called the HF transition frequency (f_T) which is the frequency at which the transistor current gain has fallen to unity in value.)

Transistor action

If the simple P-N semiconductor junction is elaborated so that it has three layers, for example to make an NPN structure, of the kind shown schematically in Figure 1.11, this can be made to act as an amplifying device, known as a junction transistor. In practice, one of the N-type layers will be only lightly doped, and this region, because of its function, will be called the collector. The other N-type region will be heavily doped, to reduce its conducting resistance, and since it is the source of the current flow it is termed the emitter. The doping levels chosen by the manufacturer for the intermediate P-type region, known as the base, will mainly be determined by the intended use of the device.

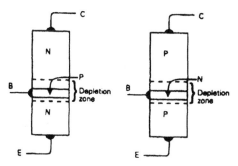

Figure 1.11
Single crystal junction transistor structures

If a reverse voltage is applied across the collector-base junction, no significant current will flow so long as the applied voltage is less than the breakdown voltage of the junction. If a small forward voltage is applied across the base-emitter junction, once again no current will flow until the forward voltage reaches the potential barrier level. However, the differences in doping types and the existence of potential differences across the junction will cause depletion zones to form, within the material, on either side of the base-emitter and base-collector junctions, so that, if the base layer is thin, the depletion zone may extend throughout its entire thickness.

Referring to the schematic drawing of Figure 1.11, if the base is made more positive, electrons will flow from the emitter into the base region, but they will not necessarily flow into the base circuit since, depending on the geometry of the junctions and the positive potential on the collector, a large proportion of the electrons which set out from the emitter will be swept straight through the base region and into the collector. This leads to a base voltage vs. collector current relationship of the kind shown in Figure 1.12, and allows the device to control a relatively large collector current by a relatively low base voltage.

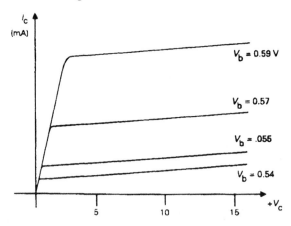

Figure 1.12
Relationship between collector current and collector voltage for small signal silicon transistor

In this respect, the bipolar junction transistor (BJT) behaves in a similar manner to the thermionic valve, except that the grid of a valve is essentially an open-circuit electrode, whereas the base region of the junction transistor has a relatively low resistance path – whose value depends on the emitter current – to the emitter. Also, while in a valve the anode current is controlled by its grid voltage, in a junction transistor it is more correct to regard the collector or emitter current as being controlled by the current flowing in the base circuit, which, as shown in Figure 1.10, has a very non-linear relationship with the applied voltage.

Junction transistors are available in a range of packages, suitable for a very wide range of collector voltages, currents and power dissipations. Provided that their dissipations are not exceeded and the integrity of the encapsulation has not failed, all modern semiconductor devices have a virtually indefinite life.
The mutual conductance of a bipolar junction transistor is very high, and can be determined from the relationship

$$g_m = I_c(q/kT)$$

where q is the charge on the electron ($1.6 ~ 10^{-19}$), k is Boltzmann's constant, and T is the absolute temperature.

This means that the mutual conductance is dependent principally upon the collector current, and for a junction temperature of 25°C the slope will be about 39S/A. For a small-signal transistor, at 1mA, a practical value of g_m could be 40mS (mA/V). This allows high stage gains with relatively low collector load resistances.

PNP and NPN transistors

A major practical advantage of the junction transistor is that, because P-type and N-type materials can be made with equal facility, it is possible to make PNP (negative collector voltage) transistors almost as easily as NPN (positive collector rail voltage) ones, by the simple substitution of differently doped materials. The availability of these two types of device is of great convenience to the innovative circuit designer. The schematic symbols used for these, and other discrete semiconductor devices, are shown in Figure 1.13.

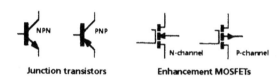

Junction transistors   Enhancement MOSFETs

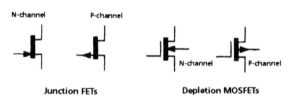

Junction FETs   Depletion MOSFETs

Figure 1.13
Transistor circuit symbols

It should be understood, however, that these complementary types of transistor are not identical, since there is a basic difference in the way in which current will pass through N- and P-type materials. In an N-type region, where there is a surplus of electrons, current flow is due to the movement of these electrons, exactly as in a metallic conductor. On the other hand, in a P-type material, the shortage of electrons results in vacancies occurring in the valency bands in places where electrons should normally be.

In the semiconductor engineer's parlance, these absent electrons are referred to as holes, and behave in much the same way as positively charged electrons. The difference, though, is in their speed of flow, because the apparent movement of a hole occurs when it is filled by an electron which has moved, in response to some electrical field, and has left a hole where it had been. This hole will, in turn, be filled by another electron which will leave yet another hole, and so on, like marbles on a solitaire board. The effect of this is that current flow in a P-type region is both slower and more noisy than in an N-type material.

A practical consideration which follows from this difference is that small-signal PNP transistors, which have an N-type base region, have a slightly lower noise factor than their equivalent NPN versions, though, in modern devices, this difference is slight. On the other hand, the greater mobility of the electrons in the relatively broad emitter and collector regions means that the HF performance of an NPN transistor will be better, other things being equal, than a PNP one. This is a factor which should be remembered in relation to power output transistors.

Junction field effect transistors

It has been seen that the presence of a reverse potential across a semiconductor junction will increase the width of the depletion zone on either side of the junction, and vice versa. Field effect transistors, normally called FETs, make use of this effect to allow a voltage applied to a reverse biased P–N junction to control the flow of current along a strip of suitably doped semiconductor material, called the channel, from an input connection – called the source – at one end of the channel to an output connection – called the drain – at the other end, as shown in Figure 1.14a. This control electrode is called the gate, and behaves like the control grid in a valve. Usually there will be two such parallel connected gate regions diffused into either side of the channel to increase the effectiveness of this electrical control action.

If there is a voltage difference between the two ends of the channel, as will normally be the case, this will have the effect of making the depletion zone somewhat unsymmetrical, as shown in Figure 1.14b. As the drain voltage is increased, in respect to the gate, the two depletion zones will move closer together until the channel is pinched off entirely, and current flow through this depleted region, still controlled by the gate voltage, is due to quantum-mechanical tunnelling. In this condition, which is

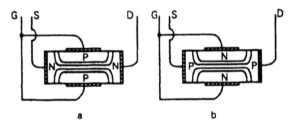

Figure 1.14a
Schematic construction of N– and P–channel junction FET

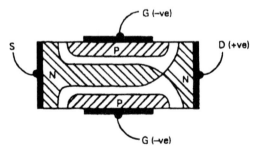

Figure 1.14b
Effect of pinch off in junction FET

that of normal operation, the drain current is almost completely independent of the drain voltage, as shown in Figure 1.15. The relationship between gate voltage and drain current is shown in Figure 1.16, and is much more linear than for a bipolar junction transistor.

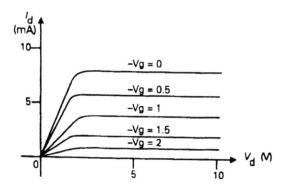

Figure 1.15
Junction FET drain current characteristics at voltages above pinch off

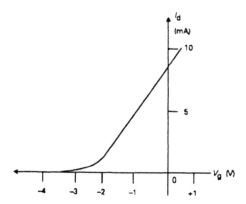

Figure 1.16
Conduction characteristics of typical junction FET

Breakdown potentials and operating characteristics

In use, the junction FET behaves very much like a thermionic valve, in that the gate is effectively an open-circuit electrode so long as the gate-source potential is reverse biased. If it becomes forward biased the gate-channel junction will be effectively the same as a forward biased junction diode. It is commonly supposed that all FETs are sensitive to, and can be destroyed by, the high voltages associated with inadvertent electrostatic discharges, but junction FETs are, in reality, no more fragile in this respect than any other small-signal diode.

By comparison with the bipolar junction transistor, FETs are relatively low voltage, low power devices, with maximum drain-gate voltages typically restricted to about 25–40V, and free-air dissipations of about 400mW. Although normal mutual conductance figures lie in the range of 2–10mS (mA/V), the in-circuit stage gain is limited by the fact that the low drain-gate breakdown voltage precludes the use of high drain load resistance values, unless very small drain currents are acceptable. Once again, the stage gain (M) can be calculated, approximately, from the equation

$$M \approx g_m R_L$$

The input and drain-gate capacitances of an FET are similar, at about 2pF, to those of a small power triode valve. Neither FETs nor junction transistors are at all microphonic unless they are mounted at the end of long, flexible leads in proximity to some charged body. Modern junction FETs are capable of very low voltage noise figures, except that their high input impedances (typically some 10^{12} ohms) mean that a high impedance input circuit can generate high values of Johnson (thermal) noise.

Insulated gate field effect transistors

These devices, known as IGFETs or MOSFETs (metal oxide/silicon field effect transistors) have become the most widely used, in the greatest range of styles, of all the field effect transistors. Their method of operation is exceedingly simple, and an attempt to make a device of this type was made by Shockley in the early 1950s. (His efforts were frustrated by his inability to obtain sufficiently pure semiconductor material.)

A simple MOSFET is shown, schematically, in Figure 1.17, and consists, in principle, of a strip of very lightly doped P-type material, into which, at either end, an N-type region has been diffused to allow electrical connections to be made to the strip. As in the case of the junction FET, these parts of the MOSFET are called the source, the drain and the channel. In manufacture the emitter contact pad (an area of vacuum deposited aluminium) is arranged to overlap both the source diffused zone and the 'P-' (a style of nomenclature which means lightly doped P-type) channel. This makes the device behave just like a very low gain NPN transistor with its base connected permanently to its emitter, a condition in which no collector current will flow.

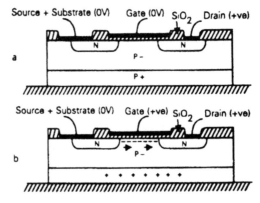

Figure 1.17
Simple insulated gate FET

If, however, a thin insulating layer (typically silica or silicon nitride in the case of a MOSFET made from silicon) is formed so that it covers the channel region, and a conducting layer (most commonly polycrystalline silicon) is formed on top of this insulating surface, then if a positive potential is applied to this conducting surface layer (also called the gate) it will induce a film of negative charges (electrons) across the gap between the source and the drain.

Since there is no difference between electrons produced by electrostatic induction and those due to any other cause, if a voltage is now applied between drain and source, a current will flow, and this current will be controlled by the gate voltage. This gate voltage vs. drain current relationship, shown in Figure 1.18, can be made very linear (leading to a low distortion in an amplifying circuit) by suitable design of the MOSFET structure (see *Siliconix Technical Article*, TA 82-3).

The style of MOSFET described above would be called an N-channel enhancement type, because the drain current, at zero gate voltage, would be exceedingly small, but would increase as the gate voltage was made more positive. Its method of construction is shown schematically in Figure 1.19.

There is, however, a further type of MOSFET called an N-channel depletion type, in which there is a significant current flow at zero gate-source voltage, but which will be

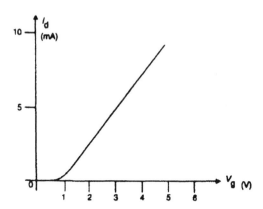

Figure 1.18
Conduction characteristics of small signal MOSFET

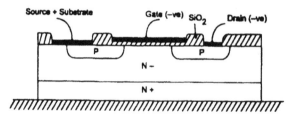

Figure 1.19
Construction of lateral P–channel enhancement MOSFET

reduced as the gate is made negative. The characteristics of this type of MOSFET are shown in Figure 1.20, and its method of construction is shown in Figure 1.21. In this, the device fabrication has been modified so that there is a thin conductive N-type layer connecting the source to the drain, underneath the gate electrode. When a positive voltage is applied between the drain and the source a current will flow, which will be reduced as the gate voltage is made more negative. In this respect the N-channel depletion MOSFET behaves in a manner which is almost identical to the thermionic triode, but without the valve's problems of microphony, fragility and loss of emission during use.

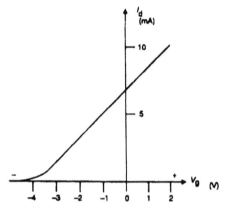

Figure 1.20
Conduction characteristics of N–channel depletion MOSFET

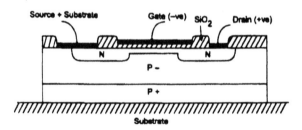

Figure 1.21
Depletion type MOSFET

A fundamental problem with the MOSFET, of whatever type, is that the channel will usually be long, and the channel conductivity will be low. In low power applications these characteristics will not be particularly inconvenient, but in higher current or higher power usage they would be undesirable. Two methods are employed to lessen these difficulties, firstly to construct the MOSFET so that the channels are vertical, when the normal fabrication methods can make them short, and secondly to arrange the layout of the device so that there are a multiplicity of channels operating in parallel.

This method of construction allows very high values of mutual conductance to be obtained: 2–5S/A for a vertically diffused power MOSFET, as compared with 2–10mS for a small-signal lateral device, having an ID of, say, 5mA. Similarly, while channel resistances of 500–3kΩ are normal for a small-signal MOSFET, on resistances (called R_{DS} on) of a fraction of an ohm are common for power devices. The penalty incurred by this multiple-channel construction is that the gate-source capacitance is very high – of the order of 1–2nF for an N-channel power MOSFET, and somewhat higher than this for an equivalent P-channel device. Because of the method of construction, the gate-drain capacitance will be smaller than this.

Because the electrostatic induction of a charge in the channel in response to the presence of an electrical potential on the gate is an almost instantaneous effect, MOSFETs have an excellent HF response, and this makes them prone to very high frequency oscillation (which may be too high to be seen on the monitor oscilloscope) if the physical layout of the circuit in which they are used is poorly arranged. D-MOS and T-MOS devices normally have a somewhat lower F_T than V-MOS or U-MOS forms, and the avoidance of parasitic oscillation is consequently rather easier to achieve in the D or T types.

Because of their very high values of F_T, it is easier to obtain low values of harmonic distortion and phase error by the use of overall negative feedback in power MOSFET based audio amplifiers than in similar circuits using bipolar junction transistors in their output stages. Also, in practice, power MOSFETs are more robust than similar bipolar junction transistors. These advantages may justify the use of the rather more costly power MOSFETs in high quality amplifier designs.

Since in RF amplifier use, the relatively high drain-gate capacitance of the MOSFET would lead to the same type of problem that exists in the triode valve, a solid state equivalent of the screened-grid valve, called a dual-gate MOSFET, is made using the type of construction shown in Figure 1.22. These transistors are suitable for use as RF amplifiers, with the drain-gate #1 capacitance reduced to about 0.01pF.

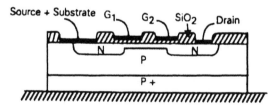

Figure 1.22
Lateral dual gate N–channel depletion MOSFET

Noise levels

Because junction transistors are fundamentally low input impedance devices, they can also offer very low levels of thermal noise, in good quality components in well–designed circuitry. Allowing for their higher input impedances, junction FETs are also available in very low noise versions. Small signal MOSFETs, however, tend to be relatively noisy, because of the mechanism by which the drain current is induced. This is particularly true for P-channel MOSFETs because the induced current carrying zone is composed of holes, whose action in tumbling over one another when carrying current leads to higher values of flicker and shot noise than in the N-channel versions 4.

Integrated circuits (ICs)

These are multiple-component devices which can be made by any type, or combination of types, of semiconductor technology. In these, a range of active devices are combined with other passive components, within a single package, to produce a complete circuit module. In general these will only require the provision of input and output connections, and either a single, or a pair of, power supply rails, to carry out some specific electronic function. These were originally manufactured to carry out various logic functions in which the very small physical sizes of the devices would allow both high operating speeds and a large number of internally interconnected gates. However, this has become the area of the most rapid growth in the whole of the electronic component field, and ICs are now made for an enormous variety of applications, from digital microprocessors to low-noise, high-gain linear operational amplifiers (op amps).

There are a number of incentives to the audio circuit designer to use linear ICs, such as op. amps, wherever their use would be satisfactory or appropriate. These are their high reliability (an IC containing a hundred transistors and a score of resistors or capacitors will be just as reliable, in general, as any single component of that type, and much more reliable than the same arrangement built up from individual resistors and transistors), their small size and their relatively low cost. In addition, since IC manufacture is an exceedingly competitive business, the internal circuitry used in the ICs will reflect the skills of some of the most gifted engineers in the field. In particular, op amps are now capable of performances, as general purpose audio gain blocks, which can hardly be matched, let alone bettered, by discrete component layouts.

Further reading

Since the principal purpose of this book is to explore the circuit designs employed in audio amplifiers, both valve based and solid-state, I have tried to restrict my analysis of semiconductor behaviour to that needed by the reader to understand what is going on. If a more comprehensive explanation of the function and construction of solid-state devices is wanted, I would recommend the reader to read Chapters 4 and 5 of my book *The Art of Linear Electronics* (Butterworth-Heinemann, 1993,).

CHAPTER 2

PASSIVE COMPONENTS

Inductors and Transformers

All conductors have inductance, and the longer the conductor, the greater this will be. Although in the frequency range of interest in audio the amount of inadvertent inductance due to things like connecting leads are not likely to be significant, they may lead to occasional, unexpected, and generally unwanted, high frequency phenomena – such as HF parasitic oscillation – which can spoil amplifier performance.

If a conductor is wound into the form of a coil, its inductance will be increased, and if this coil is wound around a core of some ferromagnetic material its inductance will be increased still further, in proportion to the permeability (μ) of the core material.

If two such coils are placed in proximity to one another, so that their magnetic fields will interact, then any change in the current flowing through one of them will cause a voltage to be induced in the other. In general, the shape of the core of a transformer will be chosen to give the highest practicable amount of electromagnetic coupling between the windings, a process which is helped by the use of high permeability core materials.

Transformers have the useful property of converting a small current flow, at a high impedance, through one of the windings, into a larger current at a lower impedance, through another winding, and vice versa. Also, because the conducting coils which form the windings on a transformer can be electrically isolated from one another, so far as static DC potentials are concerned, they allow signals or other AC voltages and currents to be transferred from one circuit to another, even when these circuits operate at widely different DC voltage levels.

Mains transformers are used in every piece of audio equipment, except those powered from batteries, to generate the DC power rails used to operate it, and output transformers are still used in every valve operated audio amplifier to couple the high impedance level of the valve output stages to the relatively low impedance levels of the loudspeaker circuit. In all of these applications, the problems are the same, and require the same measures to reduce their unwanted effects.

Problems with transformers

Ideally, the changes in magnetic flux, which occur when there is a change in the current flowing through one winding of a transformer, should interact completely with the other windings on that transformer, and not at all with any other external circuits. The use of a ferromagnetic core material – the term ferromagnetic means a type of material which will increase the inductance of a coil in which it is placed – will help to confine the magnetic flux within the transformer, and increase the effectiveness of coupling from one winding to another. The higher the magnetic permeability of the core material and the shorter the magnetic path in its core, the better this will be done.

These materials will have a relationship between magnetic flux and applied magnetising field, known as the B–H curve, of the kind shown in Figure 2.1, and the difference (X–Y) between the paths of the magnetisation and demagnetisation curves shown in this graph indicates the extent of residual magnetisation of the core, and is known as the magnetic hysteresis of the material. The value of the magnetic flux (B) for a given magnetising field (H) is a measure of the permeability of the core material. For comparison, the B–H curve for an air (or vacuum) cored system is also shown in Figure 2.1.

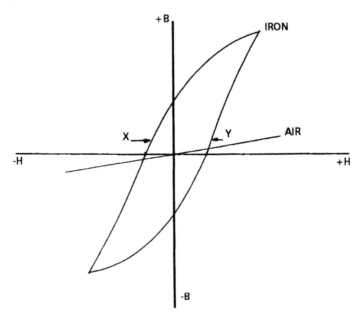

Figure 2.1
Magnetic hysteresis curve

There are two broad categories of ferromagnetic material, those which have a high value of magnetic retentivity, such as cobalt-steel, known as permanent or hard magnetic materials, and those of low retentivity or soft magnetic materials, such as iron with low carbon content.

For iron cored inductors or transformers it is essential to use only those materials which have very low levels of magnetic retentivity, or otherwise the hysteresis loss which will accompany every flux reversal in the core magnetisation will result in an absorption of input energy which will be released as heat. Also, even in low retentivity materials, the decay of the magnetic flux, on removal of the magnetic field, is not instantaneous, so there will always be an increase in core losses at higher signal frequencies.

The final major source of energy loss is that due to eddy current flow within the core. This arises because the core will generally be made of an electrically conductive metal, such as silicon steel (which has a permeability some 8000 times greater than air), and any electrically conductive paths in this will look like the short-circuited turns of a series of secondary windings. This eddy current loss can be made smaller by constructing the core from a stack of thin sheet stampings, coated with some insulating material on one or both faces, known as laminations. These can then be arranged to divide the core block at right-angles to the direction of the transformer windings and of the current flow within them. Thin laminations of high permeability core material give better transformer performance and lower losses, but they are expensive and fiddly to assemble within the former on which the coils have been wound. Like hysteresis losses, eddy current losses increase rapidly with operating frequency.

In the case of eddy current losses, T.H. Collins, *Analog Electronics Handbook*, pp. 292–293 (Prentice Hall, 1989), quotes the relationship

$$P_e = af^2B_m{}^2d^2$$

where P_e is the power loss, a is a constant, f is the operating frequency, B_m is the maximum core flux density, and d is the thickness of the laminations.

Toroidal and 'C' core constructions

These types of core are shown in Figure 2.2, and allow the practical use of very thin lamination material. In the manufacture of a C core, shown in Figure 2.2a, the core material is wound into the required shape, and then solidly locked in this shape by impregnation with epoxy resin, before being sawn into two sections to allow the windings, on a suitable former, to be set in place. If the sawn faces are ground flat, at the correct angle, the two transformer core sections can be reassembled, and clamped in place, with only a very small residual gap at their junction. In the case of the toroidal construction shown in Figure 2.2b, the core is first wound into a ring form, and impregnated, and the windings are then wound, in situ, by a suitably designed coil winding machine. Both of these constructions allow low core losses, low external radiation fields and low levels of unwanted currents induced in the secondary windings by external sources.

Toroidal types are widely used as the mains power transformers in high quality audio amplifiers because their low external 50/60Hz field greatly lessens the extent to which hum will be induced in nearby low signal level circuitry. Their high electromagnetic efficiency also provides proportionately higher power outputs, and lower winding (copper) losses than the older (but less expensive) designs based on square stacks of laminations.

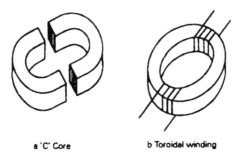

a 'C' Core   b Toroidal winding

Figure 2.2

Waveform distortion in coupling transformers

This arises both because of the non-linearity of the core magnetisation curve, and, in amplifiers, because of core saturation and input overload due to inadequate primary inductance at low frequencies. At high frequencies, the difficulty in specifying the transformer behaviour and its transformation ratio becomes much greater, because of the presence of parallel and inter-winding capacitances and inductances, as the transformer equivalent circuit shown in Figure 2.3 will make clear. In this diagram, L_P and L_S are the primary and secondary winding inductances, $L_{L,P}$ and $L_{L,S}$ are the primary and secondary leakage inductances, R_P and R_S are the primary and secondary winding resistances and C_P and C_S are the internal capacitances of the primary and secondary windings.

The inductance of any coil is determined by a number of factors, as shown in the equation

$$L = kN^2\mu a/l$$

where L is the inductance, k is a constant approximately equal to 8.3×10^{-8} when dimensions are in centimetres, μ is the permeability of the core material, N is the number of turns of the winding, a is the cross-sectional area of the core and l is the magnetic path length. By far the easiest way of increasing the inductance of any coil

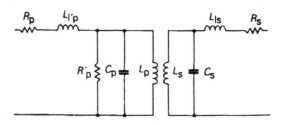

Figure 2.3
Spurious electrical components within normal two-winding transformer

is by using a ferromagnetic core material having a high permeability, such as Permalloy (μ = 20, 000 – 100, 000) or Mumetal (μ = 30, 000 – 150, 000), but all ferromagnetic core materials cause waveform distortion, and their permeability decreases with increasing operating frequency.

A dramatic demonstration of the problems caused by non-linearity in the 'B–H' characteristics of ferromagnetic materials can be given by placing a ferrite bead over the output lead between a high quality audio amplifier and its load. The immediate effect of this is that the distortion in the amplifier output will be greatly increased and changed from a smooth, low-order harmonic type to a spiky waveform rich in high-order harmonic components – distortion which disappears again if the ferrite bead is removed.

To avoid the distortion caused by the core, in those applications where relatively low values of inductance are required, as, for example, in the inductors and transformers used in low impedance loudspeaker crossover networks, air-cored coils are invariably employed, in spite of their greater physical size.

For good audio performance, in transformer coupled audio amplifiers, it is necessary to employ negative feedback (NFB) in the amplifier design, and, in view of the distortion which can arise in the loudspeaker coupling transformer, it is essential that this component shall also be included in the NFB loop. For this to be practicable without causing NFB loop instability, the low frequency performance of the transformer – principally determined by its primary inductance – and its high frequency characteristics – mainly resulting from the primary/secondary leakage inductance and the inter-winding capacitances – must also be controlled within limits set by the gain/phase relationships of the remainder of the amplifier.

An intrinsic difficulty in the use of any ferromagnetic material for the core of an inductor or transformer is that the permeability of the core material, and hence the inductance of the coil, is dependent on the magnetic flux density generated by the current flowing in the windings. This leads to the type of relationship between core inductance and AC excitation voltage shown in Figure 2.4. To ensure that the transformer performance is equally good at low signal levels as at medium levels, this

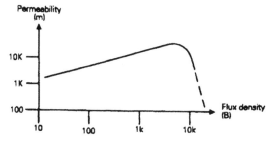

Figure 2.4
Typical relationship between permeability and flux density in silicon steel core material

minimum field inductance, known as the incremental inductance, should be the value specified. The rapid decrease in core inductance at high excitation levels, also shown in Figure 2.4, is a frequent cause of high power waveform distortion.

Where an inductor was to be used as a unit of impedance, as for example as a smoothing choke in a power supply circuit, where its high AF impedance could be used, in conjunction with a large value output capacitor, to assist in smoothing out the residual AC ripple from a simple rectifier/reservoir capacitor power supply, it was common practice to leave a deliberate air gap in the core to lessen the inductance loss due to core saturation by the DC current flow. This was also normal practice if such an inductor was used as a swinging choke, connected between the rectifier and the output of a choke input type of power supply.

An excellent analysis of the design problems which the output transformer introduces, and a practical approach to the solution of these, was given by D. T. N. Williamson (*Wireless World*, April 1947, pp. 118–121) in his article describing a design for a high quality audio amplifier. In his competitive amplifier system, P. J. Baxandall (*Wireless World*, January 1948, pp. 2–6) described a method (mainly due to C. G. Mayo of the BBC Research Dept.) which simplified the problems of overall NFB by taking the feedback voltage from a tertiary transformer winding. In practice, however, this technique gives less good results than if the NFB is derived from the LS output terminals, and segmented and interleaved windings are used, as described by Williamson, to reduce the inter-winding leakage inductance.

All in all, however, the output transformer of a valve audio amplifier is an awkward component to live with, and it could be argued that one of the major advantages inherent in low output impedance solid-state designs is that an output transformer is unnecessary. In the past, AF chokes were occasionally used as the anode loads of amplifier stages. These components would introduce similar problems, and are very rarely found in any contemporary designs.

Capacitors

These components can be divided into two main categories – polar and non-polar – and each of these capacitor types has its own set of advantages and snags. Polar capacitors are those like aluminium and tantalum electrolytics, together with some exotic ceramic-dielectric components, which are sensitive to the polarity of any direct voltage applied across them, whereas the non-polar capacitors are not. All capacitors, in their most basic form, consist of a pair of electrically conducting plates, or electrodes, separated by a gap filled with air or some other material, known as the dielectric. The differences between capacitor types relate to the nature and form of the dielectric material and the type of electrode used. In electrolytic capacitors, very large values of capacitance can be provided in relatively small package dimensions, but their actual capacitance will seldom be specified with any degree of accuracy.

Because a current must flow into, or out of, a capacitor when it is charged or discharged, if the external circuit resistance is sufficiently high that no significant current flow can occur in the external circuit, and consequently no change in the state of charge of the capacitor, then any change in the potential on one electrode of the capacitor must cause an identical potential change on the other. This provides a very useful method of transferring a voltage change from one circuit to another, when a direct connection cannot be made between them because they are not at the same DC potential. In this sense, it acts in a similar manner to an inter-stage coupling transformer, but with very much less signal distortion or sensitivity to the frequency of the signal.

By the same argument, if there is little current flow into the capacitor, then the potential across it will not change to any great extent. This is the basis of the use of a capacitor for AF bypass, or smoothing, or decoupling applications. The effectiveness of a capacitor in these uses becomes greater if the turnover frequency (f_0) due to the value of the capacitor and the resistance of the external circuit (R_{ext}) is made lower. This frequency can be found from the relationship:

$$f_0 = 1/(2\pi CR_{ext})$$

Electrolytic capacitors

The capacitance of a capacitor is related to the thickness (d) of the (insulating) dielectric layer, and the relative dielectric constant (k) of that layer (as compared with a vacuum) by the equation:

$$C = MkA/d$$

where M is some constant depending on the type of construction, and A is the area of the electrodes. The dielectric constant is a measure, analogous to that of the permeability of a magnetic core material, of the amount by which the capacitance of a capacitor will be increased by the use of that particular insulating material instead of a vacuum (or, for all practical purposes, air).

For a given electrode area the capacitance of the capacitor can be increased by making the dielectric layer thinner, and by using a dielectric with a higher value of k. The problem with making the dielectric layer thinner is that it may suffer an electrical breakdown in use, when the capacitor will cease to be usable, unless some method exists by which it can heal itself. Using a higher dielectric constant for the insulating layer between the plates is an option which depends mainly on what is available in a suitable physical form.

The electrolytic capacitor construction attempts to solve all of these problems by using an electrochemically formed layer of insulation, such as aluminium oxide on an aluminium foil electrode, or a tantalum oxide layer on a tantalum foil substrate. Not only do these oxide layers have a high dielectric constant (k = 9 for Al_2O_3, k = 26 for Ta_2O_5), but since it is arranged that some electrolyte will still be present within the capacitor body, electrolytic processes can continue, so long as there is any voltage (of the correct polarity) between the electrodes. By this mechanism, any points at which electrical breakdown has occurred will be healed automatically.

A further advantage of the electrolytic is that when the oxide film is sufficiently thick to prevent further current flow the film growth stops. This means that the dielectric thickness is never greater than is needed to arrest the flow of current at the voltage at which it was formed. Together with the high-k dielectric layer, this property means that such capacitors will offer relatively large values of capacitance in very small physical sizes. For example, a solid dielectric, epoxy resin encapsulated tantalum bead capacitor, having a capacitance of 100μF at a working voltage of 10V, can be physically only the size of a pea.

Unfortunately, the oxide insulating film in an electrolytic capacitor will tend to deteriorate during storage, or in use in applications where there is no voltage normally present between the plates, as a result of which the leakage resistance between the plates will decrease. If a voltage is then applied, a relatively large leakage current will flow, which will decrease as the dielectric layer slowly heals its weak points. Tantalum electrolytic capacitors are superior to aluminium based ones in this respect, and may even tolerate short duration DC polarity reversals, up to one or two volts.

Aluminium electrolytic capacitors dominate the larger capacitance and higher working voltage capacitor styles, with values in the range 1μF–470, 000μF at working voltages in the range of 4V–600V DC, though, of course, the larger capacitance values will usually only be available at the lower working voltages. Unfortunately, leakage currents are always present in electrolytic capacitors, and these currents are noisy. This limits their use in inter-stage coupling applications, though leakage currents are usually very low in the tantalum electrolytics. Aluminium bead capacitor types have been introduced recently which are similar in physical size and capacitance values to their more expensive tantalum based equivalents, but they are still not as good in their tolerance of polarity reversals.

A further problem in electrolytic capacitors is that there is a significant amount of resistance in series with the capacitor element, either due to the narrow internal strips of foil which are used to make connection to the wound foil electrodes, in a spiral wound capacitor, or due to the relatively poor conductivity of the electrolyte itself. This inadvertent resistive element is of importance in capacitors used for power supply applications, and is known as the equivalent series resistance or ESR, and should be as low as possible in this use.

The other measure of the lack of perfection of a capacitor, its dielectric loss, also known as dielectric hysteresis loss (roughly analogous to the core losses in a transformer), is due to the residual hysteresis losses in the dielectric when the polarity is reversed. Since this loss occurs each time the applied voltage is reversed the energy loss is roughly proportional to signal frequency, and, in an equivalent circuit, of the kind shown in Figure 2.5, would be represented by a further resistance in series with the capacitance.

In Figure 2.5b, C is the capacitance, and R_k is the dielectric loss, both of which will depend somewhat on operating frequency, voltage and temperature, R_s is the equivalent series resistance, due to internal connections, L_s is the series inductance, R_L

is the leakage resistance of the component, and the CC terms refer to various internal stray capacitances. The equivalent circuits of Figures 2.5c and 2.5d relate to electrolytic and high-k ceramic capacitors, which have further problems of reverse conduction and dielectric hysteresis, shown respectively as D and E_e in the diagrams. Although each of these spurious effects will be important in some circumstances, most of them can be ignored for most of the time.

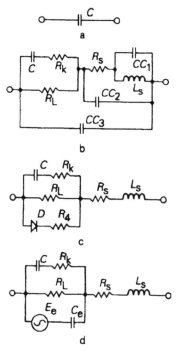

Figure 2.5
Spurious components inherent in capacitors

If there is, in use, any leakage current between the capacitor electrodes, as there will always be in electrolytic capacitors, this will appear as a resistance in parallel with the capacitor in its electrical equivalent circuit, and will increase its effective equivalent dielectric loss. This is normally measured as the tangent of the phase angle (δ) between the reactive and resistive components of an alternating current flow.

Values for tan δ, at 1kHz, range from 0.02 for an electrolytic, down to 0.00008 for an electrical grade polypropylene film. Losses increase with frequency, but, except in the case of electrolytics, will not alter much throughout the normal audio frequency range.

Non-polar capacitors

A variety of styles and dielectrics are used in these, mainly dependent on the working voltages or capacitance values required. Where small, stable, but precisely known values of capacitance are necessary, capacitors based on a stack of metallised sheets of mica or on a Swiss roll type of construction using aluminium foil interleaved with polystyrene film are preferred. These have low losses and capacitance values which are stable with time.

Where very low values of series inductance are essential, as, for example, in RF bypass applications, metallised ceramic dielectric types are preferred. These can offer either large but imprecise capacitance values in a small size, or very small but accurately defined values. All ceramic capacitors have capacitance values which are dependent on temperature, usually specified by descriptions such as N750 or P100, or simply high k (always N type), where the N or the P refer to their negative or positive temperature coefficient of capacitance, and the number refers to this temperature coefficient, in parts per million, per degree centigrade.

By far the most common types of non-polar capacitor are those made from a spiral wound sandwich of an insulating thermoplastic film dielectric and a high conductivity metal foil, usually made from high purity aluminium. A wide range of films have been used for this purpose, of which the most commonly found are polycarbonate, polyester, polypropylene, polysulphone or PTFE. The last two of these are mainly found in high temperature and military applications, where their high cost is acceptable. All wound film/foil capacitors will have some effective series inductance, because of their construction, as indicated in Figure 2.5b. This is not usually an important factor in the frequency range typical in audio applications, though some manufacturers squash their wound structure into a more slab-like form, before encapsulation, to reduce this inductance.

Self-healing properties can be built into capacitors using thin thermoplastic insulating materials by using a vacuum metallised layer deposited on the surface of the dielectric film, instead of a spiral length of metal foil. Then, in the event of an electrical puncture occurring in the film dielectric, the resulting spark discharge will burn away the metallised layer around the point of breakdown in the manner shown in Figure 2.6, and the short-circuit between the electrodes will then be cleared. This type of capacitor construction allows much thinner dielectric films to be used with safety, and consequently relatively high values of capacitance in a small package size, but, because of the thinness of the conductive layer, such capacitors will always have a high ESR, and a relatively high value of tan δ.

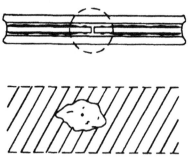

Figure 2.6
Effect of self-healing of metallized film capacitor

Waxed paper/foil capacitors were the mainstay of valve amplifier circuitry, but they always suffered from noisy leakage currents due to the ingress of moisture. Happily, these capacitors are now almost unknown in small-signal circuitry.

In audio amplifier applications, the choice will mainly lie between polystyrene (k = 2.5), polycarbonate (k = 2.8–3), polyester (k = 2.8–3.2) and polypropylene (k = 2.2) film dielectric metal foil components, depending on the requirements of the circuit application. Of these, polystyrene types will be used where low capacitance components, having stable and accurate capacitance values, are needed – as in filter circuitry or frequency response shaping networks. Polypropylene or polycarbonate capacitors would be the component of choice where it is to be used in the signal line, and low-cost polyester capacitors for other, non-critical applications.

Dielectric remanence

In certain dielectric materials the electrical dipoles within the dielectric, whose change in orientation on the application of an electric field is the major source of the dielectric constant of that material, can remain locked in their orientation even after the applied field has been removed, giving rise to an effective stored charge. In this respect, their behaviour is analogous to the persistence of orientation of the magnetic dipoles in a permanent magnet, and this effect in an insulating film is called electret formation, and is deliberately sought to provide materials for use in things such as electret microphones or headphones. Electrets occur more readily in materials having a high degree of rigidity in their structure, and are thus more readily formed in materials having an essentially crystalline structure such as biaxially oriented polyester and polypropylene films than in essentially limp and amorphous solution-cast films like those of polystyrene or polycarbonate. The ill effects, if any, of such stored charges in the audio field are not known, but their possible existence should be recognised.

A related point concerns the observations of those people who specialise in the assessment of the quality of audio systems from judgements formed during personal listening trials, and who have, as a result of these trials, declared the superiority of one type of component over another in respect of its sound quality. A particular instance of this kind concerns polypropylene film/foil capacitors, which are alleged to 'sound better' in audio systems than any other comparable type.

As it happens, I was employed for many years in the research laboratories of a large-scale manufacturer of thermoplastic films, whose product range included several types of polypropylene film, intended for use in both packaging and electrical applications. Part of my duty was to provide technical support to the manufacturers of capacitors using our electrical grade films, and we had a very well-equipped laboratory for testing films, both in sheet form, and as wound capacitors. During my work, I visited many capacitor manufacturing companies, and tested their capacitors for such things as dielectric loss, breakdown voltage and electrical remanence (stored charge).

Part of my employer's product range was a biaxially oriented polypropylene film, made from very high purity electrical grade polymer granules on machines which were kept scrupulously clean and free from potentially deleterious contaminants. This film was, of course, more expensive than the low purity, and relatively poor electrical quality, packaging grade films, made on machines principally designed to provide a high film output. The capacitor manufacturers noted this difference, but often used the cheapest film they could get, either from my own company, or from competitive suppliers with whose products I was familiar, because they said their own customers could not tell the difference, and mainly bought those capacitors which were cheapest. If there were significant differences in audio sound quality due to electrical differences, these differences would, I thought, have been exposed by those who made the pursuit of this quality their business, and would have observed that one make of polypropylene capacitor sounded better than another.

Resistors

Like capacitors, these are made in a range of types from a variety of materials, and have qualities which differ somewhat one from another.

Fixed resistors

For low power applications, where small size is generally sought, the most common construction is that of a small ceramic rod, having a suitable coefficient of thermal expansion, coated with a film of carbon, or metal, or tin oxide, or some other resistive glaze. The resistivity and the thickness of the deposited layer, applied as the first stage of manufacture, is then chosen to give a somewhat lower resistance to the composite rod – as measured between conductive metal caps attached at either end of the rod – than the target value, and a spiral groove is then cut through the resistive coating material, to lengthen the surface conductive path and increase the resistance value to the required figure. Connecting wires are then attached to the end caps, the resistor is dipped in some tough protective lacquer, and finally, the resistor value, and the tolerance (the manufacturers' permitted margin of error in the actual resistance), is printed on the body, or, alternatively, these parameters may be shown by the use of an internationally agreed colour code, as shown in Table 2.1.

For example, a 47kΩ, 5% tolerance resistor would be coded with the sequence of colour bands, starting at one end of the body, of yellow, violet, orange, gold, where the third band (orange) indicated the number of zeros (3) following the coded number 47, and the band of gold indicated a ±5% tolerance value. Where higher accuracies are required the number of colour bands is increased, so that a 47k5Ω (47, 500Ω), 1% tolerance resistor would be coded with yellow, violet, green, red and brown bands, to indicate that there would be two zeros following the number 475, and the final brown

---

In principle, since the dielectric loss factor of a capacitor is frequency dependent, and in the case of an electrolytic, also dependent on the applied voltage, it would seem probable that capacitors could introduce waveform distortion into any audio line which contained them. However, I have made many measurements of this kind of circuit, and my conclusions are that the distortion levels due to this effect are too small to be easily identifiable by conventional test procedures. Unfortunately, not finding any measurable defects does not guarantee that such performance shortcomings do not give rise to effects which may be audible to the perceptive listener.

band would denote the ±1% tolerance specification. (This colour code system has also been used to indicate the value of capacitors, measured in pF, especially radial lead polyester block types.)

Table 2.1 The resistor colour code system

Colour	Number	Tolerance
Black	0	-
Brown	1	1%
Red	2	2%
Orange	3	
Yellow	4	
Green	5	
Blue	6	
Violet	7	
Grey	8	
White	9	
Gold	(˜ 0.1)	5%
Silver	(˜ 0.01)	10%
No colour band		20%

Metal film resistors are available at dissipation ratings from 0.125 to 0.4 watts, while carbon film types cover the range 0.125–2 watts. For higher power dissipations, up to, say, 200W, wirewound resistors are available in a variety of forms.

Because semiconductor circuitry is generally associated with low voltage, low dissipation devices, it is often forgotten that there will be a maximum voltage rating as well as a maximum dissipation figure for a resistor, typically 200V for a 0.125W metal film resistor, and up to 700V for a 2W carbon film one.

All resistors, even wirewound ones, are prone to some irreversible change in their resistance value if their permitted heat dissipations are exceeded. This was a particular problem with the wax impregnated carbon-composition rod types (which were basically fired clay rods containing a proportion of graphite in their body), whose actual value would not be known, until after the firing. This value would then be moderately permanent, unless of course they were allowed to get hot in use, when they would change in their value once more. Luckily, this type of resistor – widely used in vintage audio systems – is no longer made, and only found in museum pieces.

All modern fixed value resistors from reputable manufacturers will have resistance values which are very close to their marked figures, though these can be affected by heat, so it is generally prudent to avoid heating the body excessively when soldering, or to make solder joints too close to the resistor body.

Noise

All circuits suffer from thermal or Johnson noise, proportional to the circuit resistance, its temperature and the effective measurement bandwidth. In addition, there will be a certain amount of excess or 1/f noise, as a result of the physical imperfections in the manufacture of the constituent components. There are small differences in this excess noise between the various types of modern resistors but these are usually insignificant, except in very low signal level, low-frequency circuitry. For the purist tin oxide film, metal glaze or wirewound types with welded end leads are the best. For low powers, metal film resistors are the next best choice.

Inductance

Most film type resistors are virtually non-inductive. Wirewound types are inductive, unless otherwise specified, but this is not likely to be important except in the power output stages of audio amplifiers, or in wirewound variable resistors.

Potentiometers and variable resistors

It is a common requirement that it should be possible to alter the resistance between two points in a circuit, or to select the proportion of a voltage occurring between specific points. For this purpose a range of three-terminal potentiometers are made, using many of the techniques used in the manufacture of fixed-value resistors, but with the provision of a sliding contact which allows a connection to be made to the resistor body at some chosen point along its length.

Predictably, the problem arises that any sliding contact will be less satisfactory than a fixed one in respect of its noise component or its stability of resistance value. In high quality systems the best quality potentiometers should be chosen, since the imperfections in cheap and cheerful components can have a disproportionately large effect on the system performance. In terms of quality, wirewound potentiometers are probably the best, with cermet (resistive ceramic) or conductive plastic track types the next best option. The mainstays of low cost potentiometers are those with carbon composition tracks, with either graphite rod (better) or spoon-shaped metal (less good) sliders as the means of making contact to the track.

There will always be some mechanical wear in the repeated use of any adjustable component, usually made visible or audible by the slider contact becoming intermittent, and electrically noisy in its track to slider connection, so the better quality components should always be employed in those places such as amplifier gain controls which are likely to experience the most use.

Potentiometers are also available in twin-gang types, to meet the needs of stereo systems. The tracks of these will nearly always be of relatively low quality carbon composition type, and there will often be noticeable differences in these between the proportion of the resistance tapped off by the sliders in each of the two tracks. This is even more conspicuously the case where each track of the twin-gang potentiometer is required to have a logarithmic relationship between the tapped-off resistance and the slider rotation. So, where a high standard of performance is sought, the style of potentiometer is one of the areas where particular care should be taken in the choice of components.

Switches and Electrical Contacts

These components are a matter of particular interest in audio amplifier design, because the operation of any audio system will require the interconnection of many different units, and the switching of many electrical signal paths within those units, and this can open the way to a number of potential problems. In its simplest form, the difficulty is that any conducting metallic surface which is exposed to the air will, in due course, become coated with grease, dust, and other non-conducting contaminants. The surface will also suffer from chemical degradation because of the effects of oxygen, sulphur dioxide, and other corrosive gases in the atmosphere.

If little or no electrical current passes between such contacts there may not be enough voltage across the contaminant layer to cause it to break down, so that an unsuspected high or variable resistance may be introduced in the circuit. Higher contact currents are obviously preferable from this point of view, but, on the other hand, chemical changes to the conductor surface may form thin semiconducting inter-metallic layers which are capable of passing quite high currents without disruption, but which will distort the signal voltage waveform.

If the contacts are arranged to slide over one another, as in the case of a switch, the wiping action will help to keep the contacting surfaces clean, but if there is much current flow through these, electrical arcing may occur, which will erode the contact surfaces. Overall, pure silver offers the best of the lower priced contact materials, while, if cost is unimportant, gold-plated palladium/nickel alloy contacts offer the best combination of performance with life expectancy.

Little can be done about such things as fuseholders in a loudspeaker circuit, which may have to pass substantial currents, since the fuses themselves will be relatively crude, low cost components, tin-plated at best. The only really satisfactory answer is to choose some other type of loudspeaker protection, such as a high quality, inert gas filled relay, with adequately sized, gold-plated contacts, or, alternatively, some purely electronic means of protection.

Until comparatively recently, the interconnections between components in an audio system were often regarded as being of minor importance. However, the present view – which I think to be correct – is that the means of connection are things which should be considered with equal care, in a high quality audio system, to any of the components which they join. In particular, the tracks on the printed circuit boards, or the external wiring, or the mechanical connectors which join those parts of the circuit which can pass large currents, such as the leads to loudspeaker outputs, transformers, capacitors or power supply rectifiers, should be of adequate gauge and of carefully chosen route and type if signal impairment is to be avoided.

CHAPTER 3

VOLTAGE AMPLIFIER STAGES USING VALVES

The design of valve voltage amplifier stages is essentially a simple task by comparison with much solid state linear circuit design, if only because there are not many circuit options open to the designer of valve circuits. Valves are also relatively linear in their input/output transfer characteristics, and therefore require less in the way of circuit design elaboration to improve the circuit performance and reduce any residual waveform distortion.

Circuit Configurations

For valve operated low frequency amplifier stages the only real choice is between triodes and pentodes. Both of these valve types can be employed in one or other of the five circuit arrangements shown in simplified form in Figure 3.1 – grounded cathode, grounded grid, cascode, long-tailed pair and cathode follower, sometimes called grounded anode. For simplicity, I have shown at this stage only the triode valve versions of these circuits.

In the case of a normal (grounded cathode) triode voltage amplifier stage, of the kind shown in Figure 3.1a, the major choices to be made are between the type of grid bias system which is used and the type of anode load which is employed to convert the changes in anode current, brought about by changes in the control grid voltage, into a signal voltage output.

Grid Bias Systems

A graph showing the relationship between the anode current and the grid voltage in a typical small-power triode amplifier valve (one half of an ECC83 double triode) is shown in Figure 3.2. It is required, for the lowest possible amplitude distortion in the output signal waveform, that the valve should be operated on the most linear part of its I_a/V_g characteristic curve. This implies that the grid voltage will not be driven so far negative, during the input voltage swing, that it runs into the curved region of this characteristic, near I_a cut-off, nor will it be allowed to become positive in relation to the cathode voltage, when grid current would flow, and the grid would cease to offer a high input impedance. An additional consideration is that the actual anode current should be chosen – if there is, within the desired distortion criteria, a range of current levels at which the valve could be operated equally well – to be the lowest current level within this range. This should be done to minimise the waste of energy in excessive anode dissipation, and to avoid any needless reduction of cathode life, a thing which has an inverse relationship with anode current.

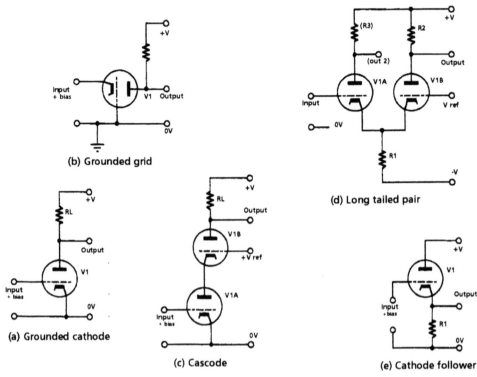

Figure 3.1
Valve gain stages

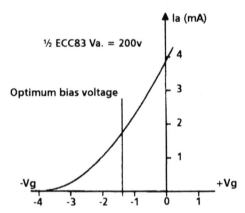

Figure 3.2
Anode current vs. grid voltage

The static working voltage applied to the grid will normally be provided in one of three ways, as shown in Figure 3.3. The simplest of these, illustrated in Figure 3.3a, takes advantage of the fact that there will be random collisions between the electrons in the 'space charge' surrounding the cathode and the control grid wires, even if the control grid is slightly negative in respect to the cathode. If a grid resistor (R1) of 2

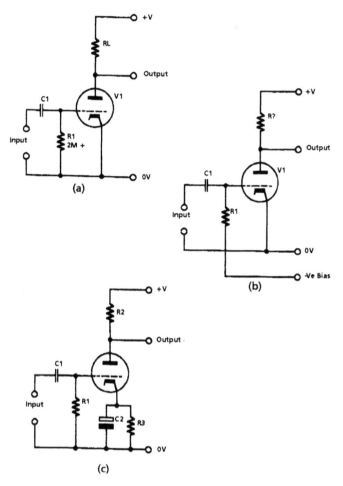

Figure 3.3
Valve biasing systems

megohms or greater is used, the grid current (perhaps 250µA) flowing through this will cause the grid to acquire a small negative voltage (perhaps 0.5V) in respect to the cathode, which will probably be adequate for a small-signal input stage. This arrangement has a number of advantages in small-signal circuitry, in addition to its simplicity, of which the major one is that it provides the highest gain figure for a given anode load impedance. The low cathode circuit impedance of this layout also assists in keeping 'hum' and noise to a low level.

This advantage is shared by the fixed bias arrangement shown in Figure 3.3b, in which any convenient value of input resistor (R1) can be used, provided that it is not so high that grid-current biasing also occurs. This layout is invariably used in directly heated (battery operated) valve circuitry, where all the cathodes (which will be directly heated filaments) will be operated from a common supply, and is occasionally used in power output stages because it will allow a slightly greater output power to be obtained for the same distortion level than would be given by a similar output stage using cathode bias. For example, Mullard quote an output from a push-pull pair of EL91 output pentodes of 4 watts, at 3.3% THD, whereas 4.8 watts could be obtained, at the same distortion figure, if fixed bias were used. A similar relationship in respect of the maximum output power levels for fixed and cathode bias arrangements exists with the KT66 and 6L6 output beam-tetrodes.

In practice, the circuit layout shown in Figure 3.3c is by far the most widely used method of generating the required level of grid bias for an indirectly heated valve. In this arrangement, a resistor (R3) is included in the cathode circuit so that the cathode current will cause the cathode voltage to rise so that it is positive to the control grid – which will be the same, in effect, as the grid being negative with reference to the cathode.

If there is no bypass capacitor across R3 the presence of the cathode circuit resistor will somewhat reduce the gain of the stage. This is due to the local negative feedback caused by the appearance at the cathode of a small AF signal, which acts to reduce the effective input signal voltage. Since the cathode resistor R3 is effectively in series with the anode circuit resistance (R2 in parallel with R_a) the feedback factor (β) will be

$$\beta = R3/(R2//R_a)$$

(in this I have used the symbol // to mean in parallel with).

If we take the case of the ECC83 triode referred to previously, at $V_a = 150V$, then $R_a = 47\ 000\Omega$, and the amplification factor (μ) will be 100. If we make R2 the same value as R_a, then the stage gain, without any cathode circuit feedback, called A^o, will be given by the relationship:

$$\text{Gain} = \mu R2/(R2 + R_a) = 100\ 47k/94k = A^o = 50$$

However, if we make R3 a 1kΩ resistor, then β will be

$$R3/23.5k = 0.042$$

The expression for gain with feedback (A`) is

$$A` = A^o/(1 + \beta A^o) = 23.8$$

which shows that the presence of an unbypassed cathode resistor will reduce the stage gain to a significant extent. The normal technique employed to lessen this drawback is to connect a capacitor (C2) in parallel with R3. At AC, the impedance of the capacitor will be:

$$Z_c = 1/(2\pi fC)$$

where f is the operating frequency. For example, a capacitor of 220μF will have an impedance of 36Ω at 20Hz, and the effect of this in parallel with the 1kΩ cathode circuit resistor will be to increase the stage gain to 46.5. This will still cause a small reduction in the LF response, and, if this is important, a larger value could be chosen for C2. At high audio frequencies the impedance of the bypass capacitor will be so low that the effect of the cathode circuit components can be neglected.

A further disadvantage in the use of cathode bias arrangements in comparison with fixed bias systems (where the cathode is tied to the 0V rail) is that the voltage developed in the cathode circuit is effectively subtracted from the anode supply voltage, and will reduce the possible undistorted anode voltage swing. This is a factor which should be considered in valve amplifier output stage design.

Cathode Bias Resistor Calculations

If the chosen amplifying device is one half of a ECC83 double triode, the required value of cathode bias resistor can be determined approximately by calculating the value of anode load resistor across which the voltage drop will place the quiescent anode voltage at about half the +V supply line potential. However, in order to determine this one must first choose a level of anode current which will place the operating point for the valve somewhere about the middle of the straight portion of the I_a/V_g relationship shown by the manufacturers (2mA at −1.4V). If the +ve supply voltage is 300V, and the anode load resistor is 75kΩ, the quiescent anode voltage will be 150V, at this value of anode current. A cathode bias resistor of 700Ω would then be required to provide the required −1.4V cathode bias. Unfortunately, the characteristic I_a/V_g curve shown for the valve is that for a +200V anode voltage rather than a +150V one, as presumed in our calculations, and this lower anode voltage will reduce the anode current flow, which will, in turn, require a slightly lower value of both cathode bias and cathode bias resistor – say 620Ω rather than 700Ω. However, the valve manufacturers will often quote a recommended value for the cathode bias resistor for a given anode load resistor and supply voltage.

Anode Load Systems

Although the resistor/capacitor circuit shown in Figure 3.4a is by far the most commonly used inter-stage coupling layout, in that it allows a wide gain bandwidth and straightforward phase/frequency relationships, the circuit stage gain is still further reduced in comparison with the simple model of Figure 3.3c because, assuming that the impedance of C3 is small enough to be neglected, the anode load resistance seen by V1 is that of R_a in parallel with both R2 and R4. It is also evident that there will be a voltage drop across R2, which will reduce the anode current for a given level of grid bias, the amplification factor (μ) and the mutual conductance (g_m) of the valve, since all of these parameters are anode voltage dependent.

The use of an LF choke, L1, as the anode circuit load, as shown in Figure 3.4b, offers an increase in stage gain in comparison with the simple resistor coupled arrangement because the DC winding resistance of the choke and the voltage drop across it will probably be low enough to be neglected. Also, if the anode current of V1 is modulated so that it swings both below and above its quiescent level, the output voltage, as seen at the anode of V1, will swing both above and below the level of the +V rail, and, for this reason, can probably give an output voltage swing rather more than double that of the circuit layout shown in Figure 3.4a.

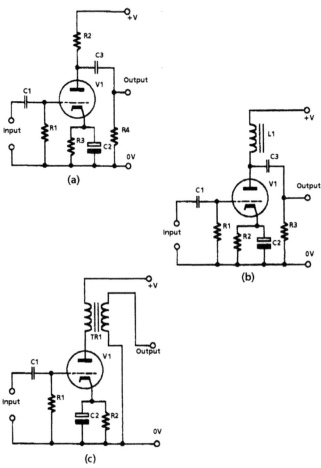

Figure 3.4
Anode load systems

The major disadvantages of the choke/capacitor inter-stage coupling system are that the phase/frequency characteristics of the choke impedance are very complex – due to the characteristics of the core material, the leakage inductance of the windings, which will either be in the form of segments or layers, and the inter-winding capacitances – and these effects will impair both the distortion characteristics and the flatness of the frequency response of the stage. This will limit the extent to which it is possible to use overall negative feedback to improve the performance of a complete amplifier.

The use of an inter-stage transformer coupling, as shown in Figure 3.4c, suffers from all the snags of choke/capacitor coupling, but will allow an even greater increase in both stage gain and possible output voltage swing in comparison with Figure 3.4a.

This made it a very popular arrangement in the early years of electronic audio amplifier designs in which it was sought to keep the number of valves used to as few as practicable. Transformer designs of that period offered voltage step-up ratios from 1:2 to 1:7 (in general, the lower step-up ratios offered better audio performance) and this could give, for example, a two valve amplifier design enough gain to drive a loudspeaker from a piezo-electric gramophone pick-up input. A particular advantage of an inter-stage coupling transformer was that, if the secondary winding was centre-tapped, the input stage could be made to drive a push-pull pair of output valves, which offered much greater output stage efficiency and output power. This topic is examined at greater detail in the chapter dealing with phase splitter arrangements.

Grounded Grid Stages

This type of circuit layout, shown in Figure 3.1b, mainly found use as an RF amplifier – in which it was effective up to very high frequencies – because the earthed control grid would act as an electrostatic screen between the output and the input circuits, which would avoid the problems due to unwanted signal feedback through the inter-electrode capacitances.

The advantages of this type of stage are that its input/output transfer characteristics are very linear when used with an input current source, or any other high impedance signal source, and that, as already noted, there is almost complete isolation between the input and output circuits. The main problem is that the input impedance (R_{in}) as seen at the cathode, is very low – approximately $1/g_m$, for an impedance in $k\Omega$ if the g_m is measured in mA/V. This could be as low as 200Ω for a valve with a mutual conductance (g_m) of 5mA/V. A further problem is that any current leakage between the heater and cathode circuits will inject hum and noise into the input signal path. Valves such as the E88CC were made especially to avoid this problem in grounded grid and cascode layouts.

Since the anode (I_a) and cathode (I_k) currents will be substantially identical, the output voltage (V_{out}) will be:

$$V_{out} = I_a R_l = I_k R_l$$

If the input impedance ($1/g_m$) is low enough in comparison with the source resistance (R_s) that it can be neglected, then the stage gain will be:

$$\frac{V_{out}}{V_{in}} \approx \frac{I_a R_l}{I_k R_s} \approx R_l/R_s$$

In other words, the grounded grid stage will only provide more than unity gain if the source impedance is less than the load impedance. The anode impedance (R_a) of the grounded grid stage is very high, approximately μR_{in}.

The Cascode Circuit

This circuit arrangement, shown in Figure 3.1c, has all of the advantages, such as linearity and input/output circuit isolation, which are associated with the grounded grid layout, but with the added benefit that the input impedance is very similar to that of the conventional grounded cathode layout of Figure 3.1a. A small disadvantage of the cascode layout is that an additional voltage supply to $V1_B$ (shown in the diagram as +V ref) is needed, and since this must allow $V1_A$ enough anode voltage (say 60V) to operate satisfactorily (the V_a of $V1_A$ will be typically Vref +1.5V) the supply voltage available to $V1_B$ will be reduced by this amount, with a consequent reduction in the possible output voltage swing in comparison with the layout of Figure 3.1a.

As is the case with the grounded grid amplifier, the anode resistance (R_a) of $V1_B$ is very high because its anode current must be the same as its cathode current, and this is simply the anode current of $V1_A$, which will be largely unaffected by the anode voltage of $V1_B$. Because of the passive operation of $V1_B$, the characteristics of the cascode stage are determined by the operation of $V1_A$, and that component of non-linear distortion in a valve amplifier which is due to the effect of variations in anode voltage on the anode current is very much reduced because the anode voltage of $V1_A$ is clamped to a fixed level by the cathode follower action of $V1_B$.

As with the grounded grid layout, a valve type designed for cascode applications should be employed, where possible, in this application to minimise the intrusions of 'hum' and noise from the heater supply line into the cathode circuit of $V1_B$. With modern technology, it would be easy to reduce the extent of this problem by operating the heater circuit of $V1_B$ from a stabilised DC supply line.

The Long-tailed Pair

The circuit layout shown in Figure 3.1d should perhaps be called a cathode coupled amplifier, but the term long-tailed pair is so widely used that other descriptions seem pedantic. As in the cascode layout, the amplifying valve operates as a grounded grid stage, with its grid, in this case, being held at a fixed voltage by a low impedance DC source (which could be the 0V rail if the tail resistor is taken to some suitable negative supply line), and with the signal input being injected into its cathode circuit.

A practical disadvantage of this circuit is that the stage gain of this arrangement is approximately half that of the simple grounded cathode layout of Figure 3.1a. This comes about because both $V1_A$ and $V1_B$ act as cathode followers in relation to the tail resistor R1. While $V1_A$ seeks to transfer the input signal present at its grid to its cathode rail, $V1_B$ seeks to hold this cathode rail voltage at a fixed and constant voltage. If both triodes are identical in their characteristics the result of these opposing endeavours will be to produce a signal voltage change at their cathodes which is half that of the input signal amplitude, at $V1_A$ grid.

A useful feature of this circuit arrangement is that, if the cathode tail resistor is sufficiently high, and if both anode circuits have identical load resistors (for simplicity, in Figure 3.1d, I have only shown one (R2) in $V1_B$'s anode circuit) the signal voltages developed at the two anodes will be very nearly equal in amplitude but in phase opposition. This allows such a stage to be used as a phase splitter preceding a push-pull output stage – this offers a great improvement in performance in this function in comparison with an inter-stage coupling transformer with a centre-tapped secondary winding. It also allows the input to be directly connected to the grid of $V1_A$, since the reference voltage at the grid of $V1_B$ can be adjusted to give the correct balance between the two valves.

The Cathode Follower

This circuit layout is shown in Figure 3.1e, and normally gives a low impedance output at a voltage gain which is slightly less than unity. The actual output voltage can be calculated by treating the circuit as a simple amplifier with a gain A^o, and 100% negative feedback (i.e. $\beta = 1$). In this mode, using half of an ECC83 double triode, the gain with feedback ($A`$) will be

$$A` = A^o/(1 + \beta A^o) = A^o/(1 + A^o)$$

If $A^o = \mu R1/(R1 + R_a)$, and $\mu = 100$, $R1 = 10k\Omega$, and $R_a = 62k\Omega$, then

$$A^o = 13.8 \text{ and } A` = 0.93$$

and the output impedance will be approximately $1/g_m \approx 600\Omega$.

An input DC bias voltage will normally be chosen so that an appropriate level of cathode current will flow and will maintain the DC operating point of the valve at a suitable position on its I_a/V_g curve. The main use of the cathode follower is to transform an input signal at a high impedance into a virtually identical signal at a low impedance, without the need for an iron cored transformer, which would introduce unwanted distortion and phase/frequency errors.

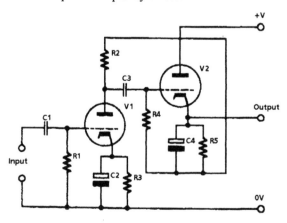

Figure 3.5
The μ follower

The μ Follower

This circuit, shown in Figure 3.5, has been recently reintroduced as a way of getting an increased stage gain from an amplifying valve (V1) by using a cathode follower (V2) as an active anode load. The cathode follower circuit gives a low output impedance, and as an active load it effectively increases the stage gain of V1. The snag is, as with all output load increasing systems, that the inter-electrode and other stray capacitances associated with the grid and cathode circuit of V2 will then have an increasingly important effect on the gain of the system, which will decrease as the signal frequency is increased. Also, an upper limit to the available gain is imposed by the shunting effect of the anode current impedance (R_a) of V1. A small-signal pentode, which will have a higher value of R_a than a triode, would be better as the input valve, V1.

Tetrodes, Beam-tetrodes and Pentodes

In the early days of valve radio receivers it was found that the triode valve was unsatisfactory as an amplifier at high frequencies because the unavoidable internal anode/grid capacitance of the valve would cause RF instability, particularly with inductive input and output circuitry.

An engineering solution to this problem was found in the introduction of a fine mesh grid, interposed between the control grid (G1) and the anode, connected to a low impedance voltage source, and held at a positive potential with respect to the cathode. This would act as an electrostatic screen between the anode and cathode and virtually eliminate the anode/grid capacitance. Since this grid was positively charged it would not significantly affect the cathode current of the valve.

The effect that this had upon the valve characteristics was to reduce the influence of the anode voltage upon the anode current. This caused a great increase both in R_a, to perhaps several megohms, and also in the magnification factor (μ), which could be as high as 5000, though because this depends so much on other things it is not often quoted by the manufacturers, and seldom used in stage gain calculations, for which the relationship $A^\circ = g_m R_1$ is normally employed rather than $A^\circ = \mu R_1 / (R_1 + R_a)$ Although the tetrode, or screened grid valve, provided an effective answer to the problems of instability in RF amplifier stages it was unsatisfactory for use as a small-signal audio amplifier stage, except at low output signal voltage levels, because of the kink in the tetrodes I_a / V_a characteristics. This kink occurs for the following reason. In any valve some secondary electrons will be emitted from the surface of the anode when high velocity electrons (from the cathode) impinge on it, especially if its surface has become contaminated with cathode metals. In a triode, this effect is unimportant because there is nowhere that such electrons can go, apart from returning to the anode. In a screened grid valve, however, these electrons will be attracted to the positively charged screening grid, which will effectively reduce the electron flow to the anode, and this will result in a downward dip in the anode current curve, as shown in Figure 3.6.

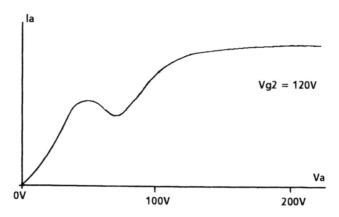

Figure 3.6
The 'tetrode kink'

Two solutions were found for this problem, of which the simplest and most widely adopted was the introduction of an open-mesh suppressor grid between the anode and the screening grid (G2). If this grid is connected to a point having a negative potential with respect to the anode – normally the suppressor grid (G3) is connected, either internally or externally, to the cathode – the decelerating potential between this electrode and the anode will cause the secondary electrons to fall back upon the anode, and the kink due to secondary emission from the anode disappears.

The second, more elegant and more expensive, solution to this problem was the so-called kinkless tetrode design, invented by the Marconi-Osram Valve Company. In this arrangement, shown in Figure 3.7, a pair of beam-confining plates, connected to the cathode, were interposed between the grids and the anode so that the electrostatic field pattern due to their shape would discourage any electron flow back towards the screening grid (G2). As a further improvement, the two sets of grid wires of the valve (G1 and G2) were aligned so that electrons from the cathode were formed into beams, and those which had passed between the control grid wires would generally fly past the screening grid wires as well, without collision.

The advantage of this structure was that it brought about a reduction in the current flow to G2, by comparison with a pentode structure, and the G2 current is just wasted energy, which contributes nothing to the design except heat. It also offered a lower distortion than an equivalent output pentode, especially when G2 was connected to the anode, to simulate an output power triode. However, beam tetrodes were more expensive than pentodes, and although there were small–signal versions of these valves, such as the KTW63 and the KTZ63, it was not generally thought that the extra cost and complexity of this construction was justified, except for power output stages where the KT66, KT77 and KT88 output beam tetrodes were the connoisseur's choice. Other output beam tetrodes were the 6V6, the 6L6 and the 807, all from the USA. Beam tetrodes seem not to have been used in mainland Europe.

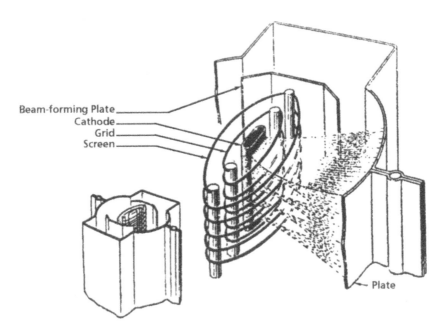

Beam-forming Plate

Cathode

Grid

Screen

Plate

Figure 3.7
Construction of beam tetrode (courtesy of RCA)

As a small-signal amplifier stage the pentode offers a mixture of advantages and drawbacks. On the credit side, it gives a much higher stage gain (say 200) in an amplifier circuit of the type shown in Figure 3.8, in comparison with that of the triode amplifier stage of the type shown in Figure 3.4a, for which a gain of 50 would be typical. On the other hand the pentode valve gives a higher level of harmonic distortion than an equivalent triode circuit, and it has a somewhat higher noise level – though this can still be very low in a well-designed valve – than a triode because of the existence of partition noise in the pentode. This defect arises because the individual electrons which collectively make up the cathode current of the valve can arrive at either the anode or G2 in a random manner, and this means that even without an input signal the anode current will fluctuate about its mean level, an effect which is heard as noise.

With suitable amendments to the circuit to provide the necessary G2 voltage, the pentode can be used in any of the circuit layouts shown in Figures 3.1 and 3.4, though unless a high stage gain is desirable a triode would be a more normal choice, if only because small-signal triodes, such as the ECC83 or the 6SN7, are very readily available as two valves having closely similar characteristics within a single envelope, and this could save space on the chassis.

A final point to be considered in the use of a pentode amplifying valve is that the screen grid (G2) current is modulated by the input signal voltage on the control grid (G1) in exactly the same manner as the anode current. It has been seen that if G2 is connected to the anode then a pentode valve will behave as though it were a triode –

with a much lower stage gain. What is not generally appreciated is that an almost identical behaviour is given if G2 is fed from the +V supply line through a sufficiently high value load resistor that the screen and anode voltages move simultaneously in the same manner. Once again, the stage gain of the valve is reduced to the level of a triode stage.

Therefore, in order to obtain the full stage gain from a pentode amplifying valve, the circuit must be chosen with an adequate value of G2 decoupling capacitor (C4 in Figure 3.8) to ensure that the G2 voltage remains at a constant level. On the other hand, it is possible to choose the values of the components associated with such an amplifier stage so that the gain will rolloff with frequency, either at the HF or LF ends of the frequency spectrum, where this gain reduction, and the phase changes associated with it, may help in stabilising an amplifier using overall negative feedback.

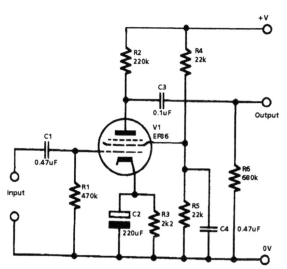

Figure 3.8
Pentode gain stage

Load Lines

If the valve manufacturers publish a full set of performance data it will be a straightforward matter to calculate the stage gain from the equation

$$A° = \mu R_l/(R_l + R_a)$$

However, the necessary data are not always published, and in this case it is necessary either to use the I_a/V_a data, if these are quoted, or to determine these for oneself. To do this it is necessary to have a fixed high voltage source, a variable low voltage (negative grid bias) source, a chosen value of load resistor, and a current meter. With this one can obtain the family of curved lines, relating anode current to anode potential for various grid bias voltages, shown in Figure 3.9 – those shown are published by Mullard Ltd for one half of an ECC83 double triode.

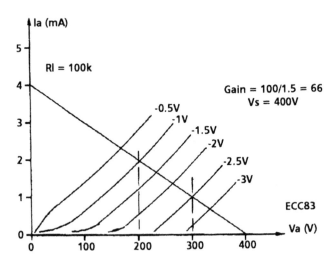

Figure 3.9
ECC83 load lines

The two points on the I_a and V_a curves which can be defined are the anode voltage with no anode current – in this case I have chosen a supply voltage of +400V – and the anode current if the valve is turned fully on, which would be 400/100k (4mA) for a 100k anode load resistor. If a straight line is drawn between these two points the intercepts on the diagonal I_a/V_g curves will be at –1V and –2.5V corresponding to anode voltages of +200V and +300V. So, with this value of load resistor a change in anode voltage of 100V will be caused by a –1.5V change in the grid bias voltage. The gain of the stage (A°), for a stage without any added cathode circuit resistance, is therefore –100/1.5 = –66 (which is the calculated stage gain for an ECC83 valve with a 'μ' value of 100 and an R_a of 50k). The same graphical process can be carried out for other values of load resistor or other values of supply voltage.

Similarly, this graphical approach can be used to determine the stage gain of a pentode, as shown in Figure 3.10. Here, the slope of the I_a/V_a curve is very much flatter than that of a triode – being almost parallel to the 0V axis – and this contributes to the higher stage gain of the valve, by comparison with a triode equivalent. However, the compression of the I_a/V_a curves at more negative grid voltages indicates clearly that the high gain pentode amplifier valve gives a higher distortion level than the lower gain triode stage. It is also obvious from Figure 3.10 that the mutual conductance (g_m) of the EF86 pentode (the change in anode current for a fixed increment in grid voltage) increases towards the higher anode current end of the anode current graph, so for maximum stage gain for a given anode load resistor the valve should be operated at the highest value of anode current which is compatible with avoidance of signal waveform clipping on the negative peaks of anode current excursions.

The triode cascode of Figure 3.1c – there is little benefit in the additional complexity of a pentode valve as either $V1_A$ or $V1_B$ – has similar gain characteristics to the pentode stage, with very much lower distortion. In this case, the simplest way of

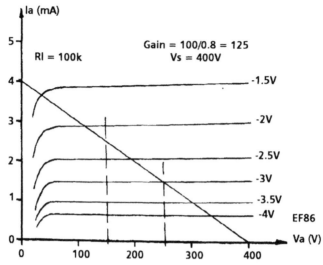

Figure 3.10
Load lines for pentode gain stage

determining the stage gain is to determine the g_m of the lower valve at the chosen anode voltage ($g_m = 1.25$ for an ECC83, operating at $V_a = 100V$, and $I_a = 0.5mA$), and then use the relationship $A^0 = g_m R_l$, so that an ECC83 cascode stage with a 100k anode load resistor would have a stage gain of 150, which is rather better than that given by the EF86. Since the voltage drop across the anode load will be 50V, a supply voltage of, say, +220V would be quite adequate, and would allow an output voltage swing approaching $100V_{p-p}$. For a higher output voltage swing, a larger value of anode load resistor and a higher +V supply line voltage would be required.

CHAPTER 4

VALVE AUDIO AMPLIFIER LAYOUTS

In its simplest form, shown in Figure 4.1, an audio amplifier consists of an input voltage amplifier stage (A) whose gain can be varied to provide the desired output signal level, an impedance converter stage (ZC) to adjust the output impedance of the amplifier to suit the load, which could be a loudspeaker, a pair of headphones, or the cutting head in a vinyl disc manufacturing machine.

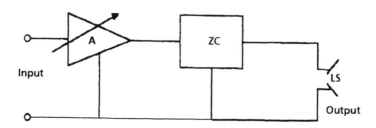

Figure 4.1
Audio amplifier block diagram

In the case of the headphones, their load impedance could be high enough for them to be driven directly by the voltage amplifier stage without a serious impedance mismatch but, with the other types of output load it will be necessary to interpose some sort of impedance conversion device – in valve operated audio systems this is most commonly an iron-cored audio frequency transformer. This is a difficult component to incorporate within a high-fidelity system, and much thought must be given both to its design and the way it is used in the circuit.

A very simple circuit layout embodying the structure outlined in Figure 4.1, using directly heated (battery operated) valves, is shown in Figure 4.2. This is the type of design which might have been built some fifty years ago by a technically minded youngster who wanted some means of driving a loudspeaker from a simple piezo-electric gramophone pick-up.

For the maximum transfer of power from an amplifier to its load it is necessary that both of these should have the same impedance, and since the anode resistance (R_a) of the output valve is of the order of 10kΩ, and the most common speech coil impedance of an inexpensive moving coil loudspeaker is 3Ω, there would be a drastic loss of available power unless some impedance converting output transformer was employed.

The primary:secondary turns ratio of this component would need to be $\sqrt{(10k/3)} =$ 58:1. It is difficult to design transformers having such high turns ratios without losses in performance and, in consequence, when higher audio quality was required, the LS manufacturers responded by making loudspeaker drive units with higher impedance speech coils. Before the advent of transistor operated audio amplifiers the most common LS driver impedance was 15Ω.

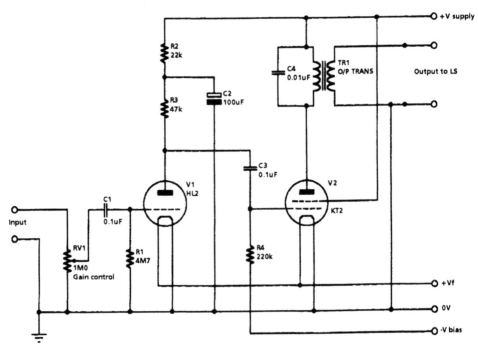

Figure 4.2
Simple valve amplifier

With regard to the amplifier design shown, the input stage (V1) uses a simple directly heated triode, with grid-current bias developed across the 4M7Ω grid resistor, R1, as shown in Figure 3.3a. This is resistor/capacitor coupled to V2, a small-power beam tetrode or pentode, operated with fixed bias derived from an external DC voltage source.

Because both V1 and V2 will contribute some distortion to the signal – in the case of V1, this will mainly be second harmonic, but in the output valve (V2) there will also be substantial third harmonic component – the output signal will sound somewhat shrill, due to the presence of these spurious high signal frequency components in the output. The simplest and most commonly adopted remedy for this defect was to connect a capacitor (C4) across the primary of the output transformer (TR1) to roll-off the high frequency response of the amplifier as a whole, to give it the required mellow sound. The HT line decoupling capacitor (C2) serves to reduce the amount of spurious and distorted audio signal, present on the +V supply line, which will be added to the wanted signal present on V2 grid. An amplifier of this type would have an output power of, perhaps, 0.5 watts, a bandwidth, mainly depending on the quality of the output transformer, which could be 150Hz– 6kHz, ±6dB, and a harmonic distortion, at 1kHz and 0.4 watts output, of 10%.

The amplifier shown in Figure 4.2 uses a circuit of the kind which would allow operation from batteries, and it was accepted that such designs would have a low output power, and a relatively poor performance in respect of its audio quality: this was the price paid for the low current drain on its power source. If, however, the amplifier was to be powered from an AC mains supply, the constraints imposed by the need to keep the total current demand low no longer applied, and this gave the circuit designer much greater freedom. The other consideration in the progress towards higher audio power outputs was the type of output stage layout, in that this influenced the output stage efficiency, as examined below.

Single-ended vs. Push–pull Operation

These two options are shown schematically in Figure 4.3, in which Q1 and Q2 are notional amplifier blocks, simplified to the extent that they are only considered as being either open-circuit (O/C) or short-circuit (S/C), but with some internal resistance, shown as R3 (or R1 in the case of Figure 4.3b). I have also adopted the convention that the current flow into the load resistor (R2) is deemed to be positive when the amplifier circuit is feeding current (as a current source) into the load, and negative when the amplifier is acting as a current sink and drawing current from R2 and its associated power supply. I have also labelled the voltage at the junction of these

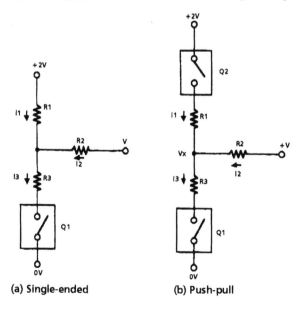

(a) Single-ended   (b) Push-pull

Figure 4.3
Output arrangements

three resistors as V_x. The efficiency of the system can be considered as related to the extent of the change in the current through R2 brought about by the change from o/c to s/c in Q1 or Q2.

If we consider first the single-ended layout of Figure 4.3a, when Q1 is O/C, the current flow into R2 is only through R1. and $i_2 = V/(R1 + R2)$. If, however, Q1 is short-circuited, S/C, then, from inspection,

$$i_3 = i_1 + i_2 \tag{1}$$

$$\text{but } i_2 = (V - V_x)/R2 \tag{2}$$

$$i_1 = (2V - V_x)/R1 \tag{3}$$

$$\text{and } i_3 = V_x/R3 \tag{4}$$

from (1), (2) and (3) we have

$$i_3 = (2V - V_x)/R1 + (V - V_x)/R2 \tag{5}$$

but it has been seen from (4) that $i_3 = V_x/R3$

$$\text{therefore } (2V - V_x)/R1 + (V - V_x)/R2 = V_x/R3 \tag{6}$$

If we insert the actual values for R1, R2 and R3, we can discover the difference in output current flow in the load resistor (R2) between the o/c and s/c conditions of Q1. For example, if all resistors are 10Ω in value, when Q1 is s/c, V_x will be equal to V, and there will be no current flow in R2 and the change on making Q1 o/c will be (V/20)A. If R1 and R2 are 10Ω in value and R3 is 5Ω, then the current flow in R2, when Q1 is o/c, will still be (V/20)A, whereas when Q1 is s/c, the current will be (−0.25V/10)A and the change in current will be (3V/40)A. By comparison, for the push-pull system of Figure 4.3b, the change in current through R2, when this is 10Ω and both R1 and R3 are 5Ω in value, on the alteration in the conducting states of Q1 and Q2, will be (2V/15)A, which is nearly twice as large.

The increase in available output power from similar output valves when operated at the same V+ line voltage in a push-pull, rather than in a single-ended, layout is the major advantage of this arrangement, although if the output devices have similar distortion characteristics, and the output transformer is well made, the even harmonic distortion components will tend to cancel. Also, the magnetisation of the core of the output transformer due to the valve anode currents flowing in the two halves of the primary winding will be substantially reduced, because the induced fields will be in opposition.

In addition, an increase in the drive voltage to the grids of the output valves, provided that it is not large enough to drive them into grid current, will, by reducing their equivalent series resistance (R1, R3 in the calculations above), increase the available output power, whereas in the single-ended layout the dynamic drive current cannot be increased beyond twice the quiescent level without running into waveform clipping. However, there are other problems, which are discussed below.

Phase Splitters

In order to drive a pair of output valves in push-pull it is necessary to generate a pair of AC control grid drive voltages which are equal in magnitude but in phase opposition. The simplest way of doing this is to use a transformer as the anode load for an amplifier stage, as shown in Figure 3.4c, but with a centre-tapped secondary winding.

I have shown in Figure 4.4 a typical centre-tapped transformer coupled 20 watt audio amplifier, of the kind which would have been common in the period spanning the late 1930s to early 1940s. Because there are two coupling transformers in the signal path from the input and the LS output, which would cause substantial phase shifts at the ends of the audio spectrum, it would be impractical to try to clean up the amplifier's relatively poor performance by applying overall negative feedback (from the LS output to V1 cathode) to the system.

Some local negative feedback from anodes to grids in V2 and V3 is applied by way of C4/R4 and C6/R7 in an attempt to reduce the third, and other odd-order, harmonic distortion components generated by the output valves. Since the designer expected that the output sound quality could still be somewhat shrill, a pair of 0.05μF capacitors, C7 and C8, have been added across the two halves of the output transformer primary windings to reduce the high frequency performance. These would also have the effect of lessening the tendency of the output valves to flash over if the amplifier was driven into an open-circuited LS load – an endemic problem in designs without the benefit of overall negative feedback to stabilise the output voltage.

The anode voltage decoupling circuit (R3,C3), shown in Figure 4.4, is essential to prevent the spurious signal voltages from the +V supply line to the output valves being introduced to the output valve grid circuits. This would, in the absence of the supply line decoupling circuit, cause the amplifier to oscillate continuously at some low frequency – a problem which was called motor-boating, from the sound produced in the loudspeakers.

Various circuit arrangements have been proposed as a means of generating a pair of low distortion, low phase shift, push-pull drive voltages. Of these, the phase inverter circuit of Figure 4.5 is the simplest, but does not offer a very high quality performance. It is, in principle, a bad thing to attenuate and then to amplify again, as is done in this arrangement, because this simply adds just another increment of waveform distortion, due to V2, to that contributed by V1.

A much more satisfactory arrangement is that shown in Figure 4.6, in which V2 is operated as an anode follower, which like the cathode follower, employs 100% negative feedback, though in this case derived from the anode. This stage contributes very little waveform distortion. Also, because both valves operate as normal amplifier stages, the available voltage from either output point will be largely unaffected by the operation of the circuit. An additional advantage over the circuit shown in Figure 4.5 is that the two antiphase output voltages are equal in magnitude, without the need to adjust the preset gain control, RV1.

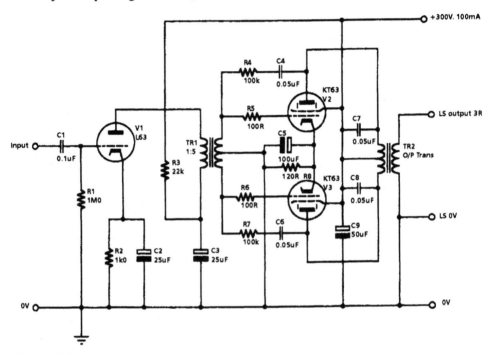

Figure 4.4
Simple 20 watt amplifier

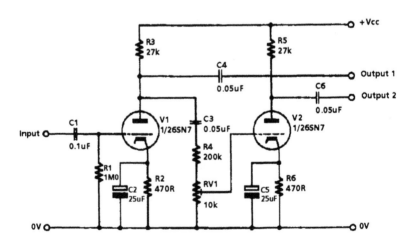

Figure 4.5
Simple phase inverter

Another satisfactory circuit is that based on the long-tailed pair layout, in which, provided that the tail resistor is large in relation to the cathode source resistance ($1/g_m$), the two antiphase anode currents will be closely similar in magnitude. The advantage of this circuit is that it can be direct coupled (i.e. without the need for a DC blocking coupling capacitor) to the output of the preceding stage, and this minimises circuit phase shifts, especially at the LF end of the pass–band. By comparison with the two preceding phase-splitter circuits, it has the disadvantage that the available AC output swing, at either anode, is greatly reduced by the fact that the cathode voltages of V2 and V3 are considerably positive in relation to the 0V line, and this will almost certainly require an additional amplifier stage between its output and the input of any succeeding triode or beam-tetrode output stage.

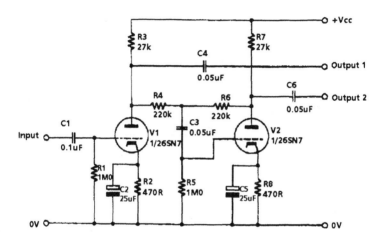

Figure 4.6
Floating paraphase circuit

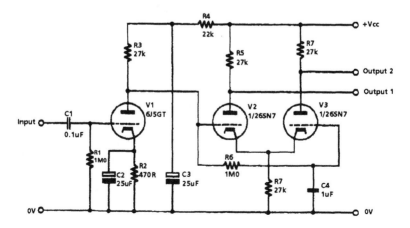

Figure 4.7
Long-tailed pair circuit

This disadvantage is shared by the circuit layout shown in Figure 4.8, in which a direct coupled triode amplifier is operated with identical value resistive loads in both its anode and cathode circuits. Because of the very high level of negative feedback due to the cathode resistor both the distortion and the unwanted phase shifts introduced by this stage are very low. Significantly, this was the type of phase splitter adopted by D.T.N. Williamson in his classic 15 watt audio amplifier design.

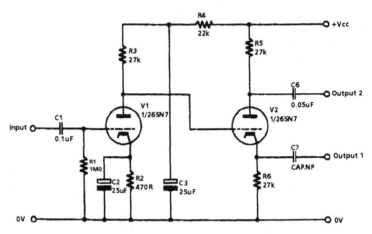

Figure 4.8
Split load phase splitter

Output Stages

The basic choice of output valve will lie between a triode, a beam-tetrode or a pentode. If large output powers are required – say, in excess of 2 watts – triode output valves are unsuitable because the physical spacing between control grid and anode must be small, and the grid mesh must be relatively widely spaced, in order to achieve a low anode current resistance and a high practicable anode current level. This closely packed type of construction will lead to the almost complete stripping of the space charge from the region between the cathode and the grid. Experience shows that the life expectancy of cathodes operated under such conditions is short, and the only way by which this problem can be avoided is by the use of a directly heated (filament type) cathode construction, which is much more prolific as a source of electrons, and this leads to other difficulties such as hum intrusion from the AC heater supplies, and the awkwardness of arranging cathode bias systems.

So, if it is required to use a triode output stage, at anything greater than the 50mA anode current obtainable from a parallel connected 6N7 double triode (the 6SN7 has a smaller envelope and, in consequence, a lower permissible anode dissipation), a directly heated valve such as the, now long obsolete, 6B4 or PX25 would need to be found. Therefore, in practice, the choice for output valves will be between output beam-tetrodes or pentodes. Although a fairly close simulation of a triode characteristic can be obtained in both of these valve types if the anode and G2 are connected together, this approach works better with a beam-tetrode than a power output pentode because the presence of the suppressor grid in the pentode somewhat disturbs the anode current flow.

The required grid drive voltage for typical pentode or beam-tetrode output valves, at $V_a = 300V$, will be in the range $20\text{--}50V_{p-p}$ for each output valve, and whether or not the valve is triode connected has little effect on this requirement. Triode connection does, however, greatly affect the anode current impedance, which is reduced, in the case of the KT88, from $12k\Omega$ to 670Ω, and the need for a lower turns ratio greatly simplifies the design of the necessary, load-matching, output transformer with low half-primary to half-primary and primary to secondary leakage inductances.

Output (Load-matching) Transformer

This component is probably the most important factor in determining the quality of the sound given by a valve operated audio amplifier and the performance of this component is influenced by a number of factors, both mechanical and electrical, which will become of critical importance if an attempt is made to apply negative feedback (NFB) over the whole amplifier. However, for a low power system, such as might be used as a headphone amplifier, it is possible to make a quite decent sounding system without the need for much in the way of exotic components, circuit complexity or very high quality output impedance-matching transformers, and I have sketched out in Figure 4.9 a typical circuit for a two valve, 1W, headphone amplifier based on a pair of 6SN7s or equivalents.

In this design the input pair of valves acts as a floating paraphase phase-splitter circuit, which provides the drive for the output valves. Since the cathode currents from the two

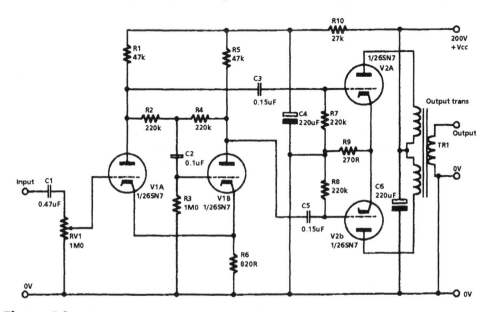

Figure 4.9
Simple headphone amplifier

input valves are substantially identical, but opposite in phase, it is unnecessary to provide a cathode bypass capacitor to avoid loss of stage gain. Also, since this cathode resistor is common to both valves, it assists in reducing any differences between the two output signals, because the arrangement acts, in part, as a long-tailed pair circuit. Since the total harmonic distortion from a push-pull pair of triodes will probably be less than 0.5%, and will decrease as the output power is reduced, provided a reasonable quality output transformer is used, I have not included any overall NFB, and this avoids any likely instability problems. To match the output impedances of V2A and V2B to a notional load impedance of 100Ω a transformer turns ratio, from total primary to secondary, of 12:1 is required.

In more ambitious systems, in which NFB is used to improve the performance of the amplifier and reduce the distortion introduced by the output transformer, much more care is needed in the design of the circuit. In particular, the phase shifts in the signal which are introduced by the output transformer become very important if a voltage is to be derived from its output, and fed back in antiphase to the input of the amplifier, in that to avoid instability the total phase angle within the feedback loop must not exceed 180° at any frequency at which the loop gain is greater than unity. This requirement can be met by both limiting the amount of NFB which is applied, which would, of course, limit its effectiveness, and by controlling the gain/frequency characteristics of the system.

Although there are a number of factors which determine the phase shifts within the transformer, the two most important are the inductance of the primary winding and the leakage inductance between primary and secondary, and a simple analysis of this problem, based on an idealised, loss-free transformer, can be made by reference to Figure 4.10. In this, R1 is the effective input resistance seen by the transformer, made up of the anode current resistance of the valve, in parallel with the effective load resistance, and L1 is the inductance of the transformer primary winding. When the signal frequency is lowered, a frequency will be reached at which there will be an attenuation of 3dB and a phase shift of 60°. This will occur when $R1 = j\omega L1$, where ω is the frequency in radians/sec.

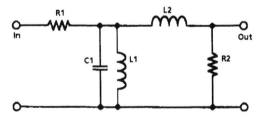

Figure 4.10
Equivalent circuits of idealised coupling transformer

R2 is the secondary load resistance, which is the sum of the resistance reflected through the transformer and the anode resistance, and L2 is the primary leakage inductance – a term which denotes the lack of total inductive coupling between the primary and secondary windings – which behaves like an inductance between the output and the load, and introduces an attenuation, and associated phase shift, at the HF end of the passband. The HF –3dB gain point, at which the phase shift will be 60°, will occur at a frequency at which $R2 = jwL2$.

To see what these figures mean, consider the case of a 15Ω resistive load, driven by a triode connected KT66 which has an anode current resistance of 1000 ohms. Let us assume that, in order to achieve a low anode current distortion figure, it has been decided to provide an anode load of 5000 ohms. The turns ratio required will be $\sqrt{(5000/15)} = 18.25:1$ and the effective input resistance (R1) due to the output load reflected through the transformer will be 833 ohms. If it is decided that the transformer shall have an LF –3dB point at 10Hz, then the primary inductance would need to be $833/2\pi10 = 833/62.8 = 13.26H$. If it is also decided that the HF –3dB point is to be 50kHz, then the leakage inductance must be $833/2\pi50\ 000 = 2.7mH$. The interesting feature here is that if an output pentode is used, which has a much higher value of R_a than a triode, not only will a higher primary inductance be required, but the leakage inductance can also be higher for the same HF phase error.

Unfortunately, there are a number of other factors which affect the performance of the transformer. The first of these is the dependence of the permeability of the core material on the magnetising flux density, as I have shown in Figure 4.11. Since the current through the windings in any audio application is continually changing, so therefore is the permeability – and with it the winding inductances and the phase errors introduced into the feedback loop. Williamson urged that, for good LF stability, the value of permeability, μ, for low values of B should be used for primary inductance calculations.

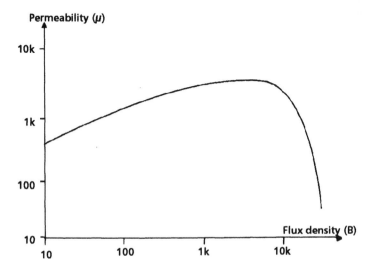

Figure 4.11
Magnetisation curve

Secondly, this change in inductance, as a function of current in the windings, is a source of transformer waveform distortion, as are – especially at high frequencies – the magnetic hysteresis of the core material and the eddy current losses in the core. These problems are exacerbated by the inevitable DC resistance of the windings, and provide another reason, in addition to that of improved efficiency, for keeping the winding resistance as low as possible.

The third problem is that the permeability of the core material falls dramatically, as seen in Figure 4.11, if the magnetisation force exceeds some effective core saturation level. This means that the cross-sectional area of the core (and the size and weight of the transformer) must be adequate if a distortion generating collapse in the transformer output voltage is not to occur at high signal levels. The calculations here are essentially the same as those made to determine the minimum turns per volt figure permissible for the windings of a power transformer (see Linsley Hood, J., *The Art of Linear Electronics*, Butterworth-Heinemann, 1993, pp. 29–31).

In practical terms, the requirements of high primary inductance and low leakage inductance are conflicting, and require that the primary winding is divided into a number of sections between which portions of the secondary winding are interleaved. Williamson proposed that eight secondary segments should be placed in the gaps left between ten primary windings. This increases the stray capacitance, C1, across the primary winding and between primary and secondary coils. However, the HF phase errors introduced by these will probably be unimportant within the design frequency spectrum.

Effect of Output Load Impedance

This is yet another area in which there is a conflict in design requirements, between output power and output stage distortion. In Figure 4.12a, I have shown the output power given for 1% and 2% THD values by a push-pull pair of U/L connected KT88s

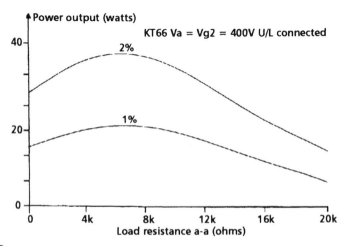

Figure 4.12a
Power output vs. THD

in relation to the anode to anode load impedance chosen by the designer. This data is by courtesy of the GEC (*Audio Amplifier Design*, 1957, p. 41). Since the distortion can also alter in its form as a function of load impedance, I have shown in Figure 4.12b the way these circuit characteristics change as the load resistance changes. The figures given for an single-ended 6AK6 output pentode are due to Langford-Smith (*Radio Designers Handbook*, 4th Edition, 1954, p. 566).

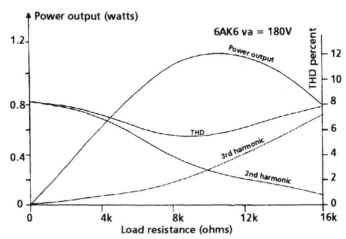

Figure 4.12b
Power output curve

Available Output Power

The power available from an audio amplifier, for a given THD figure, is an important aspect of the design. Although there are a number of factors which will influence this, such as the maximum permitted anode voltage or the maximum allowable cathode

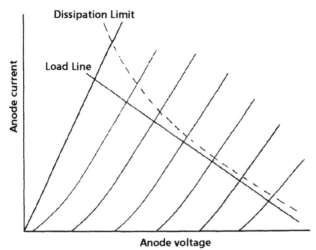

Figure 4.13
Anode dissipation curve

current, the first of these which must be considered is the permissible thermal dissipation of the anode of the valve. These limiting values will be quoted in the manufacturers' handbooks, and from these it is possible to draw a graph of the kind I have shown in Figure 4.13, where the maximum permitted combinations of anode current and anode voltage result in the curved (dashed) line indicate the dissipation limits for the valve, and the load line for its particular operating conditions can then be superimposed on this graph to confirm that the proposed working conditions will be within these thermal limits.

CHAPTER 5

NEGATIVE FEEDBACK

The concept of feeding back into its input some part of the output signal from an amplifying block was invented, like so many other useful ideas in the field of electronics, by Major Edwin Armstrong, of the US Army Air Corps. His initial intention was to use positive (i.e. signal level enhancing) feedback (PFB) to make the electrical oscillators which were required as signal sources for radio broadcasts, but it was soon discovered that the feeding back of an antiphase signal – signal diminishing or negative feedback (NFB) – from the amplifier output into its input circuit also conveyed some valuable advantages. In particular, simple NFB can reduce the extent of any waveform distortion or noise introduced by the amplifier block, and can help to make the frequency response of the system more uniform.

I describe this usage as simple NFB because there are many other circuit possibilities using NFB in which the feedback is used to cause specific modifications to the frequency response characteristics of a circuit block, as, for example, in the construction of frequency response correction circuitry or notch or bandpass filters. These will be considered later.

The effects caused by the normal – flat frequency response – type of voltage feedback are most easily explored by reference to the simple schematic layout shown in Figure 5.1. In this a circuit block having separate, and isolated, input and output connections, m-n, p-q, and a gain of A^0, is connected so that a proportion of the output voltage is fed back, in series with the input signal, E_i, through a network, FB, having a gain of β. If we define the signal actually appearing between the input points m-n as e_i, and

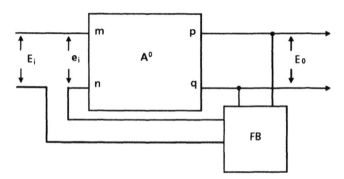

Figure 5.1
Layout of feedback system

that appearing across the output points p-q as E_o, then the stage gain of the amplifying block, A^o, is

$$A^o = E_o/e_i \qquad (1)$$

which is called the open loop gain – i.e. the gain before any feedback is employed. However, if the electrical network, FB, connected between the output and input of this gain block causes a proportion β of the output signal to be inserted in series with the input voltage E_i, then the actual input applied to the gain block will be changed, so that e_i will actually be $E_i + \beta E_o$, or

$$E_i = e_i - \beta E_o \qquad (2)$$

These equations allow us to determine the effect of feedback on the stage gain, bearing in mind that in the absence of any feedback signal, $E_i = e_i$.

The closed loop stage gain, i.e. that when feedback is applied, is normally called A', and is defined by the equation

$$A' = E_o/E_i \qquad (3)$$

Combining equations (2) and (3)

$$A' = E_o/(e_i - \beta E_o), \quad \text{but} \quad e_i = E_o/A_0$$

therefore

$$A' = \frac{E_o}{E_o/A^o - \beta E_o} = \frac{1}{(1/A^o - \beta)} \quad \text{or} \quad \frac{A^o}{1 - \beta A^o}$$

from this $A'/A^o = 1/(1 - \beta A^o)$ $\qquad (4)$

One of the effects which can be seen immediately from equation (4) is that, if the feedback is positive, then if βA^o is unity, the stage gain will be infinite. This consideration is important for two reasons:

- that if there is some mechanism in an electrical circuit which limits the gain once a certain output amplitude is reached, then the use of positive feedback can be made to provide a continuous and stable amplitude output signal from the circuit without any input signal being required. In general, some frequency determining network will be employed to ensure that this output oscillation occurs at some desired and specified frequency;

- that if there is some characteristic of the layout or the components used in the gain block which tends to cause a phase shift in the fed back signal, so that, although the feedback factor β is intended to be negative, it is actually positive at some frequency, then continuous oscillation may occur at that frequency if the closed loop gain reaches unity. Not only that, but the stage gain will tend to increase, in any amplifier employing negative feedback, where the phase shift of the fed back signal causes this to be positive, even though the amount of the feedback may not be enough to cause oscillation.

There are a number of advantages and disadvantages which are inherent in the use of feedback in an amplifying system, and it is essential, in any design, that its designer should be aware of the latent problems as well as the required benefits.

Benefits of Negative Feedback

If the feedback network (FB) is a phase inverting one, so that β is negative, then the gain equation will be

$$A^{\cdot} = A^{\circ}/(1 + \beta A^{\circ}) \tag{5}$$

In this case the closed loop gain will decrease as the amount of feedback is increased, up to the point at which βA° is so much greater than 1 that the equation approximates to

$$A^{\cdot} = A^{\circ}/\beta A^{\circ} \quad \text{or} \quad A^{\cdot} = 1/\beta \tag{6}$$

This relationship is of particular interest where A° is very high, because in these cases the gain with feedback, $A^{\cdot}$, is almost completely determined by the attenuation ratio, β, of the feedback network, so that if β is 1/10 then the gain $A^{\cdot}$ will be 10, and if β is 1/100, then the stage gain will be 100, and so on. This is a valuable feature in the case of valve amplifiers, or other circuits where the performance of the components may be expected to deteriorate with time, in that it will sustain the gain and other qualities of the circuit in spite of component deterioration.

Some further valuable features can be predicted, particularly in the limit case where $A^{\cdot}$ tends towards $1/\beta$, for example that, if the feedback network has an attenuation which is independent of frequency, then the action of the feedback will be such as to make the amplifier gain also independent of frequency. Also, if the feedback network is distortion free, then, if the gain (A°) is high enough the output will also be distortion free. A logical deduction from this is that, for those cases where A° is less than infinite in value, the distortion (D°) of the output signal will be reduced in direct proportion as the value of βA° increases, giving the relationship

$$D^{\cdot} = D^{\circ}/(1 + \beta A^{\circ}) \tag{7}$$

where D° is the distortion without feedback. However, this relationship makes the assumption, not always found in practice, that the open loop gain (A°) is still sufficiently high, and negative, at the frequencies of the various higher harmonic components which go to make up the total distortion. A similar prediction can be made about the reduction of 'hum' and noise introduced by the amplifier block, provided that the feedback network is itself hum and noise free. This feature is particularly valuable in valve operated audio amplifiers, since the poor smoothing of the HT+ supply lines, and the likely induction of hum, from external AC fields, into the windings of the output transformer means that the output signal from a simple amplifier is typically not as free from hum as would be desired.

Another advantage of NFB is that it can improve the accuracy of rendition of transient signals, in that any difference between the input signal, E_i, and that which is fed back, βE_o, will generate an error voltage which will be amplified by A^o and subtracted from the signal output. All in all, an amplifier block employing negative feedback, where this is applied correctly in a properly designed system, will be substantially superior to that of a similar system without NFB. The fact that claims are sometimes made that the performance of an amplifier without NFB is superior to that of one using this technique seems to be due principally to the lack of design skills of those in respect of whose designs this claim is made.

Stability Problems

It has already been noted that phase shifts within the amplifier block, or the feedback network employed, can lessen the stability of the circuit to which NFB is applied. Various methods have been employed to demonstrate this effect, but one of the simplest, and most easy to understand, is that of the Bode plot shown in Figure 5.2. In this, the closed loop gain and the phase angle of the fed back signal are plotted, simultaneously, against frequency in the range of interest.

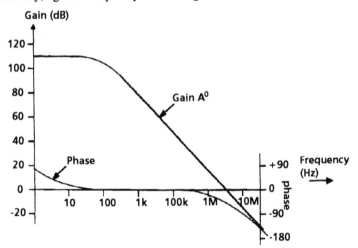

Figure 5.2
The Bode plot

Stability in the system requires that the gain of the system should not exceed unity (0dB) at any frequency at which the phase of the fed back signal reaches or exceeds −180°. This requirement arises because the initial phase of a negative feedback signal will be −180°, so that an added phase shift of 180° will change the phase of the signal fed back to −360° (= 0°), and if the gain equals or exceeds unity the system will oscillate.

Two methods are used to describe the stability of an amplifier using NFB; the gain margin, which refers to the extent to which the loop gain is less than unity by the time the phase shift has become −180°, and the phase margin, which refers to the extent to which the phase angle of the fed back signal is less than −180° by the time the loop gain has decreased to unity. In general the negative signs are omitted from the description, so a 20dB gain margin would imply that the loop gain would be − 20dB (0.1) at the frequency at which the loop phase shift was −180°. Alternatively, a 30° phase margin would imply that the loop phase shift had only reached −150° at the frequency at which the loop gain was unity.

The problem in audio is that the phase shift within an amplifier will be influenced by the characteristics of the load applied to it, as well as by the signal frequency, and probably also the amplitude of the signal. Unfortunately, all of these things are continually varying in any real-life audio system, so any system without an adequate margin of stability may be flickering into and out of the regions of near or actual instability for the whole of the time, and this will lead to the modification of the signal waveshape, due to instantaneous changes of gain, or, at the worst, to the generation of bursts of oscillation, superimposed on the wanted signal. It is difficult to generate suitable input test signals to reveal this problem, and it is also hard to simulate in the laboratory appropriate awkward loads to allow examination of these effects.

Understandably, badly designed amplifiers will give unpleasant results, and these results may worsen when NFB is applied, but the answer is not to use reduced amounts of NFB or no NFB at all, but to design them so that they have greater loop stability margins, especially when used with reactive loads. It is inconvenient that this is a region in which the overall stability in operation of an amplifier is often in conflict with the attempts of the designer to achieve the lowest practical level of total harmonic distortion, when measured with a steady state signal on a purely resistive load. For this purpose, a large amount of NFB is often used, and the design of the circuit is then tailored to sustain the loop gain at a high level up to the frequency at which it must fall rapidly to reach 0dB before the loop phase approaches −180°. Circuits designed like this, though stable, often have very small gain and phase margins,

Series and Shunt NFB

It is convenient in this analysis to use the conventional operational amplifier symbol, shown as 'A' in Figure 5.3, which denotes a circuit block with a very high internal gain, and two input points, labelled − and +, to indicate either inverting or non-inverting operation of the amplifier. It is assumed in this convention that the amplifier block will have a very low output impedance, and that both of the input points will have a very high input impedance. It is also assumed that the amplifier acts only in a differential mode, in which it will amplify only those signal voltages which exist between its two inputs, and will ignore any potential difference which may exist, jointly, between both of the inputs and the common 0V rail.

In the circuit of Figure 5.3a, the input signal is connected to the non-inverting input, while the feedback signal is taken to the inverting input through the resistor network

R1 and R2. From the calculations shown above, the gain will be 1/β, or

$$A' = (R1 + R2)/R2 \qquad (8)$$

and, if the gain block has a very high impedance at its two inputs, the input impedance of the circuit of Figure 5.3a will be that of the resistor R_{in}. This method of connection is described as series feedback, because the feedback voltage is applied at the amplifier input in series with the input voltage, since only the difference between the two input voltages will be amplified.

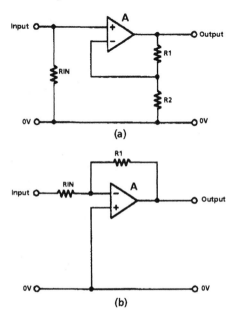

(a)

(b)

Figure 5.3
Feedback systems

There is a limit, in theory, to the performance which can be obtained from such a feedback connected amplifier block, because there will be some failure, however small, in the rejection of the voltage present on both input terminals – the so-called common mode voltage. In the better transistor op amps, the extent of the common mode rejection may be very good, particularly at the lower end of their operating frequency range, but in a simple valve operated differential amplifier, such as the long-tailed pair layout shown in Figure 3.1d, the cancellation of common mode signals may not be very good, and distortion due, for example, to the non-linearity of the I_a/V_g characteristics of the two valves, will not be completely cancelled.

In the circuit layout of Figure 5.3b, both the input and the feedback voltages are applied to the inverting input of the amplifier block, in parallel, or shunt, and this mode of operation is known as shunt feedback. The gain of the stage is

$$A' = R_1/R_{in} \qquad (9)$$

and the input impedance is that due to R_{in}, because the inverting input will appear to be what is termed a virtual earth, a concept which will be explained later. Since both inputs are connected to the same point there is no common mode error.

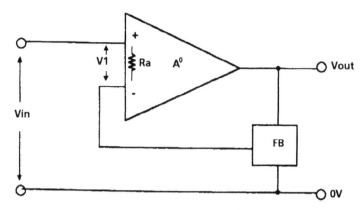

Figure 5.4
Series feedback system

The Effect of NFB on Input and Output Impedances

The aspect of the series NFB system can be explored with reference to Figure 5.4. In this, the amplifier block has a gain of A^o, β is the proportion of the signal which is fed back by the network, FB, V_1 is the actual voltage which appears across the input of the gain block, and R_a is the input resistance of the gain block. It is apparent that

$$V_{out} = A^oV_1$$

but $\qquad V_{in} = \beta V_{out} + V_1$

therefore $V_{in} = \beta A^oV_1 + V_1$

$$= V_1(A^o + 1)$$

The actual input resistance (R_{in}) of the system is:

$$R_{in} = V_{in}/i_{in}, \text{ and } i_{in} = V_1/R_a$$

so, since $V_{in} = V_1(A^o\beta + 1)$, then, in the case of a series feedback system,

$$R_{in} = \frac{V1(A^o\beta + 1)}{V_1/R_a} = (1 + A^o\beta)R_a \qquad (10)$$

which is to say that the input impedance is effectively increased in proportion to the amount of feedback employed.

In the case of the shunt feedback layout shown in Figure 5.5, if we define the circuit input current as i_1, the amplifier input current as i_2, and the current through the feedback resistor, R2, as i_3, then:

$$i_1 = i_2 + i_3 \tag{11}$$

and, if the current amplification of the amplifier is A^o, then

$$i_3 = A^o i_2 \quad \text{or} \quad i_2 = i_3/A^o \tag{12}$$

Combining these two equations we get

$$i_1 = i_3/A^o + i_3 = i_3(1 + 1/A^o)$$

from which $\quad i_1 = i_3(A^o + 1)/A^o$

so, by inversion,

$$i_3 = i_1 A^o/(A^o + 1) \tag{13}$$

If $V_{in} = i_1 R_1$ and $V_{out} = -i_3 R_2$, because the gain block is a phase inverting amplifier, then we can determine the circuit gain with feedback (A') which is

$$A' = \frac{V_{out}}{V_{in}} = \frac{R_2}{R_1} \frac{A^o}{A^o + 1} \tag{14}$$

which, if A^o is high enough, will approximate to:

$$A' = -R_2/R_1$$

The impedance at point 'X' of an amplifier employing shunt connected NFB (which I have called R_a) can be determined, approximately, by the following argument.

Let us postulate that the gain block is an ideal amplifier in which neither input draws any current, so that $i_1 = i_3$. However, because it is an inverting amplifier, the voltage across R2 will be $(A^o + 1)V_{in}$, and $i_1 = \{(A^o + 1)V_{in}\}/R_2$

so,

$$R_{in} = \frac{V_{in}}{i_{in}} = \frac{V_{in}}{\{(A^o + 1)V_{in}\}/R_2}$$

which simplifies to
$$R_{in} = R_2/(A^o + 1) \tag{15}$$

This can be very low indeed if A^o is high enough – much lower in practice than the input impedance (R_a) of any likely amplifier block – so that neglecting the amplifier input current, i_2, will not significantly affect the calculated value for R_{in}. The very low input impedance at the inverting input of any amplifier using shunt feedback has given rise to the term virtual earth to describe its characteristics, and allows the circuit arrangement shown in Figure 5.6 to be used as a summing amplifier or signal mixer with very little interaction between the various inputs.

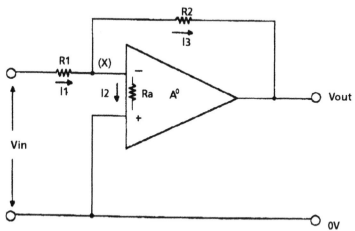

Figure 5.5
Shunt feedback layout

To summarise: the series feedback system offers a very high input impedance, which will increase as the open loop gain of the system is increased. By contrast with this, the shunt feedback layout has a very low input impedance, which will decrease as the open loop gain of the amplifier is increased. In normal shunt FB voltage amplifier circuits, there will be an input resistor, such as R1 in Figure 5.5, and this will be effectively the input resistance of the circuit.

A further important difference between shunt and series connected NFB is that if the feedback limb, between output and inverting input, has zero impedance then, in a shunt feedback system, the gain of the stage will be zero, whereas, in a series feedback

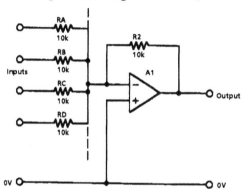

Figure 5.6
Virtual earth mixer system

system, the gain will be unity. This difference is important in those circuit layouts, considered below, in which the feedback network contains capacitors, whose impedance will decrease, and will ultimately tend to zero as the signal frequency is increased.

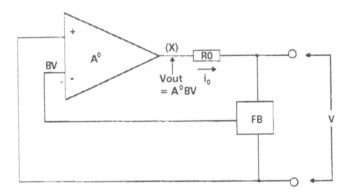

Figure 5.7
Arrangement to allow calculation of output impedance

Influence of NFB on Output Impedance

In both shunt and series connected NFB circuits the application of NFB will lower the output impedance of the gain block, as measured at the point from which the NFB is taken. The magnitude of this effect can be calculated by considering the case of an amplifying block with a gain A^o, and an effective output resistance, without feedback, of R_o, and has a feedback signal returned to its inverting input via the network, FB, which has an attenuation of β.

If a signal voltage, V, is applied to its output, this will produce a voltage βV at the inverting input of the amplifier, and this will, in turn, generate a voltage of $-A^o \beta V$ at point 'X'. The voltage across R_0 (V_0) is given by

$$V_0 = V - V_{out} = \{V - (-A^o \beta V)\} = V(1 + A^o \beta)$$

and the current through R_0 (i_0) will be

$$i_0 = V_0/R_0 = V(1 + A^o \beta)/R_0$$

so the output impedance with feedback, R'_0, which is V/i_0, will be

$$R'_0 = \frac{V}{V(1 + A^o \beta)/R_0} = \frac{R_0}{(1 + A^o \beta)} \tag{16}$$

and this effect of lowering the output impedance is the same for both feedback systems.

Effect of NFB on the Harmonic Components of the Signal

A frequently overlooked consideration, in the application of NFB to improve the linearity of non-linear amplifier circuits, is that, in small amounts, NFB can both worsen the total harmonic distortion and modify the distribution of the harmonic components within the distorted signal. The reason for this is simple – that if, for example, an amplifier block introduces second and third harmonic distortion, these harmonics will be present in the feedback signal to the amplifier, and will then be distorted further by the amplifier to produce new fourth, sixth and ninth harmonic components, not present in the original distorted output. So, while the second and third harmonic components present in the feedback will tend, following amplification, to cancel the original distortion components, the higher order distortions will look like a new input signal, which will pass through the system, to be amplified and distorted in their turn.

A practical example of this problem was shown graphically by P. J. Baxandall (*Wireless World*, December 1978, pp. 53–56), whose results are reproduced in the graph shown as Figure 5.8. These figures were determined by measurement on a simple FET amplifier stage, for various levels of applied NFB. For a system using more than 15dB of NFB, the reduction in distortion is as would have been predicted from the simple mathematical model, but for small amounts of NFB, up to, say, 10dB (equivalent to a reduction in gain of 3.3˘), the distortion could actually worsen due to the increase in the magnitude of higher order components. This problem is more likely to arise in valve operated than in transistor based amplifier systems, partly because it is difficult to get high stage gains from single stage valve circuits, and partly because of the difficulty in applying NFB around multiple stage amplifiers – especially those which employ transformer coupling.

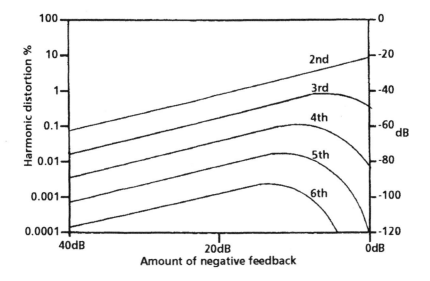

Figure 5.8
Effect of NFB on harmonic distortion

Practical Valve NFB Systems

Negative feedback circuit layouts are widely used as frequency response shaping systems, using resistors and capacitors in networks chosen for their impedance vs. frequency characteristics. Such a feedback path, whose input is derived from the anode, can be returned either to the valve grid or to its cathode. By far the most popular arrangement for this type of system is that shown schematically in Figure 5.9. In this type of layout a single valve, usually a pentode, is operated under conditions chosen to give a high stage gain, and with shunt connected NFB taken through some frequency dependent network, (Z) from the anode to the valve control grid.

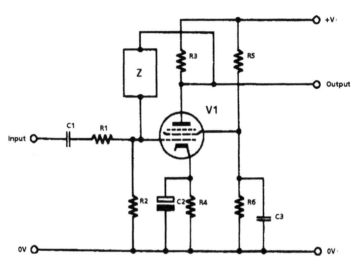

Figure 5.9
Simple NFB arrangement

This arrangement would give an approximate stage gain of Z/R1. In early valve preamplifier circuitry both the components forming the feedback network Z and the input network, R1,C1, would be controlled by a multi-position wafer switch to allow a range of different gain/frequency options to be selected to suit radio, or gramophone pick-up, or tape recorder, or microphone inputs.

Although the alternative series NFB connection is possible, this usually means that the feedback signal must be introduced into the cathode of the amplifier valve, which is a very low impedance point, and this presents problems.

A very successful series feedback circuit arrangement, nicknamed the ring of three for obvious reasons, and much favoured in physics labs for use as a high gain, linear, wide bandwidth gain module is shown in outline form in Figure 5.10. With suitable components this could give a gain of up to several hundred, with very little waveform distortion and with gain/bandwidths flat from, say, 20Hz up to 20MHz or more. The approximate gain of this circuit, which will be constant over a wide frequency band, will be (R7 + R8)/R8.

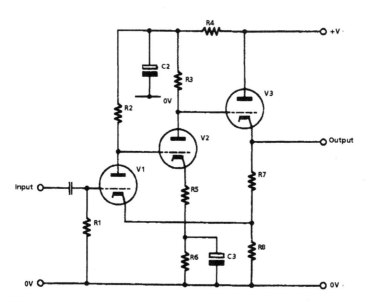

Figure 5.10
The 'ring of three' feedback circuit

The Baxandall Negative Feedback Tone Control System

Contemporary fashion decrees that the user of Hi-Fi equipment should not attempt to modify the frequency response of his or her equipment to increase or diminish the relative loudness of any part of the audio frequency spectrum with respect to the remainder, and that, instead, the best available flat frequency response equipment should be purchased, and the tonal rendering given by this should then be left alone. However, in more practical, and perhaps more realistic, times it was accepted that imperfections could indeed exist, and most of the better audio systems provided some form of 'tone control' to allow the user to compensate for any tonal shortcomings in the bass or treble output of his or her audio equipment.

Of all the circuits suggested for this purpose, by far the most popular – and adopted by audio designers on a world-wide basis – was the method proposed by P.J. Baxandall (*Wireless World*, October 1952, pp. 402–405) of which I have shown the basic layout in Figure 5.11. I regret that, among the purists, even this excellent circuit is thought likely to impair the quality of the final sound – even when the controls are adjusted so that the circuit gives a flat frequency response – and component arrangements are favoured in which any desired adjustment to the frequency response is made only by the use of passive components (such as resistors or capacitors) in the belief that these cannot introduce any unwanted worsening of the sound quality.

Unfortunately, this argument is fallacious in that, for any passive tone control to work, it must introduce an overall attenuation in its flat frequency response mode which is at least equal to the desired boost in gain at the desired frequency (treble or bass) relative to its gain in the mid-range. So, in order for the overall gain of the system, including

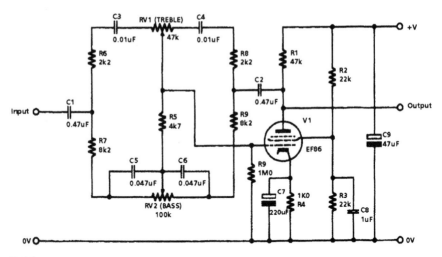

Figure 5.11
Baxandall NFB Tone Control

the control circuit, to be the same as it had been before the circuit was inserted, an additional (distortion introducing) gain stage must now be used to make up for the loss of gain due to the passive circuitry. Unfortunately, this amplifier stage is likely to introduce more distortion than that of the equivalent stage within the feedback system, because any distortion introduced by this will be reduced by the internal NFB within the circuit. This argument applies equally to such circuit as filters or fixed frequency response adjustment networks (i.e. for gramophone PU output equalisation) as it does to tone controls.

To understand the operation of the Baxandall circuit it is convenient to postulate that the two tone controls are set at their mid positions (in Baxandall's original circuit, the potentiometer RV2 was centre-tapped, and this tap was connected to the 0V line, but in normal practice, this refinement is omitted, since the circuit works quite well without it). In this case, the input limb of the circuit (to V1 grid, which acts as a virtual earth because the circuit employs shunt connected NFB) is R6,C3 and half RV1 in parallel with R7,C5 and half RV2. The feedback limb is, similarly, R8,C4 and half RV1 in parallel with R9,C8 and half RV2. Clearly, at all audio frequencies, these two limbs are equal in impedance and the circuit therefore has a gain of unity.

If the slider of RV1 is moved to the left, then, at (treble) frequencies at which C3 and C4 have a low impedance, the impedance of the left hand (input) limb of this network will be less than that of the right hand (feedback) one, and the circuit gain will be increased up to a limit value of

Treble gain (or attenuation) $= (2200 + 47\,000)/2200 = 22.3$

which is just over 26dB. Conversely, if the slider of RV1 is moved to the right, the gain at treble frequencies will be decreased to an ultimate attenuation of −26dB.

A similar argument can be used to determine the limit increase in gain, or the limit value of attenuation at those very low audio frequencies at which the impedance of C5 and C6 is much higher than the resistance of RV2. In this case, the bass gain will be

Bass gain (or attenuation) $= 108\ 200/8200 = 13.2$

which is just over 22dB. Intermediate settings of RV1 and RV2 will provide intermediate levels of lift and cut of either treble or bass, and these adjustments can be made simultaneously, and with very little cross influence. Also, the amplifier valve will have an open loop gain of about 50, and this will reduce the closed loop distortion (at mid-frequency range) to about 0.01% at 6V rms output.

By comparison with this, a similar gain stage following a passive tone control layout, with NFB used to adjust its gain to, say, 25, will have a distortion (at mid-frequency range), of about 0.2% at 6V rms output. The drawback in both cases is that any treble boost, even in a distortion-free circuit, will enhance the amount of harmonics of the signal as it increases the treble gain, relative to that of the waveform fundamental frequency. On the other hand, treble attenuation will reduce the THD. An analogous and very obvious phenomenon, within the experience of any user of vinyl gramophone record discs, is that any use of bass boost will exaggerate the audible, low-frequency rumble noise caused by imperfections in the turntable bearings.

Frequency Response Equalisation

In the early years of the manufacture of 78rpm emery-loaded shellac gramophone record discs a variety of modifications were made to the frequency response of the recorded signal in the hope that the inverse frequency response, used on replay to recover a flat frequency output from the recorded signal, would lessen the surface noise of the disc, and increase the practicable recording levels. By the time the vinyl (LP) gramophone record was introduced, in 1951, universal agreement had been obtained on the type of record and replay characteristics. The agreed standard, for both 45rpm and 33rpm discs, is shown in Figure 5.12, and is generally known as the RIAA response curve because it was proposed by a consultative committee from the Radio Industries Association of America.

A typical gramophone record replay circuit, of which the characteristics have been tailored to give an RIAA type replay frequency response, is shown in Figure 5.13. This circuit gives a stage gain of 10 at 1kHz, and since the input signal is likely to be small, a cascode layout is chosen to give the lowest noise practicable from a shunt feedback circuit.

As seen from Figure 5.12 it is required to generate a gain/frequency response curve which rises, below 1kHz, at which frequency the relative gain is 0dB (unity) to +3dB at 500.5Hz, and +17dB at 50.05Hz. Above 1kHz the relative gain is required to decrease from −3dB at 2121.5Hz to −20dB at 21.21kHz. This type of response curve is produced by the two parallel RC groupings shown as R2C2 and R3C3 in the circuit drawing, since the overall gain will be determined by the relative impedance ratios of the two limbs R2C2 and R3C3 relative to R1C1. The method by which the required component values can be calculated is given in my book *The Art of Linear Electronics* (Linsley Hood, J. (1994), pp. 172–174, Butterworth-Heinemann), as is also a rather fuller account (in Chapter 7) of the use of NFB in audio systems.

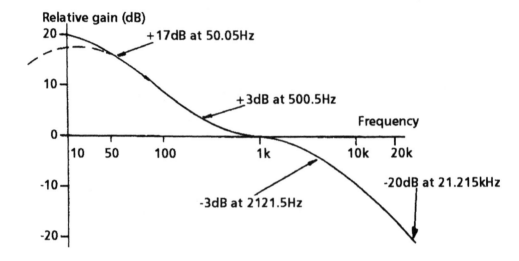

Figure 5.12
RIAA equalised frequency response

Filter Circuits Based on NFB

D.T.N. Williamson, in part of his celebrated series of articles describing a very high quality audio amplifier and preamplifier (*Wireless World*, October/November 1949), described two circuits in which he used a combination of a frequency selective feedback arrangement and a parallel-T notch filter to provide a steep cut high pass rumble filter and an equally steep cut, but switched frequency low pass filter which would serve to reduce the amount of audible high frequency hiss resulting from the emery powder loading of the shellac body of the contemporary 78rpm records.

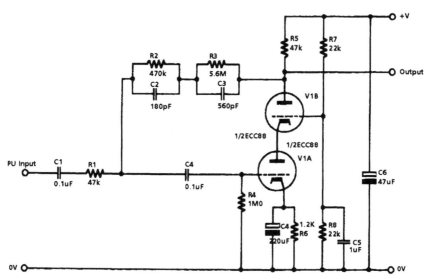

Figure 5.13
Gramophone (RIAA) pick-up input stage

The basic layouts of these two filter types are shown in Figures 5.14a and 5.14b, and the frequency response of the final circuitry incorporating these is shown in Figure 5.15. The method of operation of both of these filter circuits is the same: if NFB is applied around a circuit in which there are two RC low pass networks operating at a similar roll-off frequency, the action of the NFB will be to flatten the frequency response up to some turnover frequency determined by the RC networks, beyond which the output will fall at an ultimate rate of –12dB/octave.

In the case of the low pass circuit shown in Figure 5.14a, the two RC networks are formed by R4,C1 and R7,C5. By introducing a parallel-T notch filter (R2,R3,C4 and C2,C3,R5) into the signal path, and then taking the NFB point from the output of this, the overall flatness of the frequency response is maintained, but the slope of the HF roll-off beyond the turnover frequency is increased by the notch circuit to about –40dB/octave.

The action of the high pass filter circuit of Figure 5.14b is essentially similar, except that only the first of the two CR networks (C1,R10 and C4,R7) is within the feedback loop, and the T network is bridged to reduce the ultimate attenuation rate to about –30dB/octave, which Williamson judged to be sufficient to remove rumble noises due to turntable bearing imperfections.

The detailed design of these circuits is given in a reprint of his original articles (Williamson, D.T.N., *Wireless World*, August 1949–May 1952. Reprinted in 1990 by Audio Amateur Publications, Inc., Peterborough, NH, USA).

An alternative example of valve operated preamplifier design, the Brimar SP55 stereo preamp, is shown in Figure 5.16. This dates from the middle 1950s and is quoted in

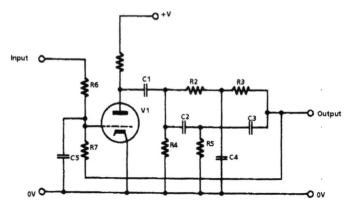

Figure 5.14a
Low pass filter

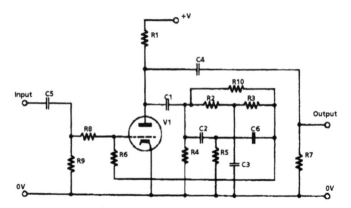

Figure 5.14b
High pass filter

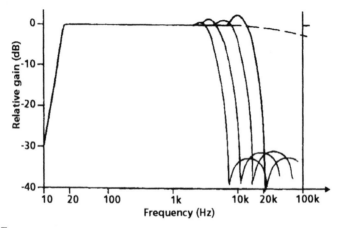

Figure 5.15
Filter characteristics

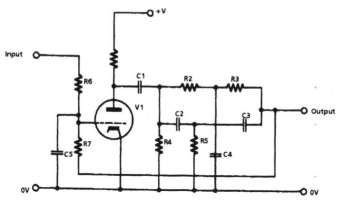

Figure 5.14a
Low pass filter

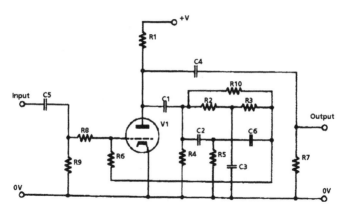

Figure 5.14b
High pass filter

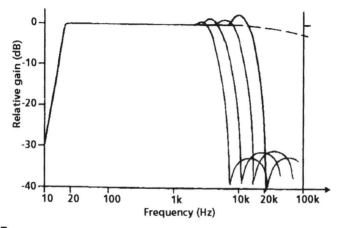

Figure 5.15
Filter characteristics

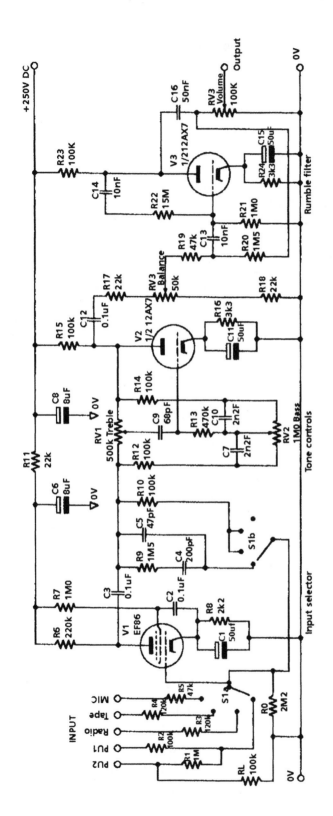

Figure 5.16
The Brimar SP55 preamp

the *Brimar Valve and Teletube Manual,* Volume 9. This is a typical, and well-engineered, example of circuit design at a period when the situation with regard to gramophone record replay facilities had stabilised with the more or less universal adoption of 33 $^1/_3$ rpm LPs and 45 rpm EPs in place of the easily broken 78rpm shellac discs. Because hiss on record replay was no longer a problem, steep-cut treble filters of the type Williamson proposed were no longer much needed, and were seldom offered. Also, with the new vinyl discs came an agreed (RIAA standard) replay frequency compensation characteristic and an effective choice between high performance and expensive electromagnetic pick-up cartridges, and inexpensive and popular piezo-electric systems, usually based on a piezo-electric ceramic element.

For both of these cartridge types a standard load of 100k was adopted (contemporary usage would favour a load resistance of 47k) which was supposed to constrain the ceramic cartridge into an electromagnetic style of performance, for which, with both types of input, the frequency response would be corrected by the network R9,C4,C5 in the anode-grid feedback path of V1. For all the other inputs provided, V1 would act as a flat frequency response stage whose gain was determined by the input and feedback networks, with the exception of the microphone input where no input stage NFB was employed.

This allowed V1 to have a stage gain of about 200, and gave input sensitivities, for maximum output:

Mag. p.u, 8mV, Ceramic p.u, 60mV, Radio/Tape recorder, 100mV and microphone, 2.5mV.

The resistance network connected around V2 implements a Baxandall type feedback tone control circuit, for which the quoted performance is:

Treble lift/cut, ±13dB at 10kHz, ref. 1kHz.
Bass lift/cut, ±16dB at 50Hz, ref. 1kHz.

The final stage provides an input channel balance facility coupled with a permanently connected rumble filter, with a flat frequency response down to the cut-off point beyond which the gain would fall rapidly, with an ultimate attenuation rate of −18dB/octave. This filter characteristic was generated by NFB through the two feedback loops, C14,R22 and C16,R20,C13, connected between anode and grid of V3. A roll-off frequency of 28Hz was adopted in this design since it had by now become generally accepted that there was little below 30Hz which was either part of the musical content of the recording or capable of reproduction by anything but the very best of LS systems.

The design figure for signal output, with tone and balance controls set at mid-position, and the volume control set at its maximum output position, was 500mV, at which the distortion figure would be less than 0.05%. The quoted signal/noise ratio for the disc inputs was better than −55dB, and for the radio and tape inputs, better than −65dB.

Although modern design practice would allow some improvement on these signal/noise figures, it is in this respect that the biggest difference exists between valve and solid state technology, for which a disc signal/noise ratio of, say, −85dB would be normal, and a flat frequency response auxiliary input position (such as provided for a CD replay input) should offer signal/noise ratios of the order of −100dB.

CHAPTER 6

VALVE OPERATED AUDIO POWER AMPLIFIERS

There was only a limited range of circuit options open to the audio amplifier designer before the widespread availability and adoption of solid state semiconductor components, and for this reason the variety of commercially successful designs evolved for valve operated audio power amplifiers was also fairly limited. In this chapter I have mainly concerned myself with those circuits produced by UK manufacturers, or those designs offered both by the valve manufacturers' laboratories and by independent designers for home construction.

By the early 1950s the expected standard of performance of a high quality audio amplifier was that it should have an output power of at least 10 watts, that it should have a full power bandwidth of at least 20Hz–20kHz, and that its harmonic distortion, at 1kHz and full output power, would be better than 0.2%. In order to achieve this sort of performance, it was necessary, in practice, that overall (output to input) NFB should be used. It was not normally feasible to do this with the simple transformer coupled layout shown in Figure 4.4 because the inevitable phase shifts, introduced within the feedback loop by the two coupling transformers, would make the amplifier oscillate continuously, except where the designer had chosen to use very small – and probably unhelpful – feedback levels.

In the case of the designs using only one transformer (that for coupling the output valves to the loudspeaker) it was possible to design the circuit to allow useful amounts of NFB to be employed – a value for βA° of 20–26dB was a typical figure – provided that the output transformer was well designed, and this meant that, for the designer, the extent of the choices available to him were in the type of input amplifier stage (triode, pentode, cascode or long-tailed pair), the type of phase splitter (split load, floating paraphase, long-tailed pair or simple inverter), the type of output valves,(pentode or beam-tetrode, straight or triode connected), whether an additional amplifier stage (usually a push-pull pair of triodes) between the phase splitter and the output stage was judged to be necessary and, finally, in the case of pentode or beam-tetrode output valves, whether the output transformer had primary tapping points to which the screened grids could be connected – the so-called ultra-linear or distributed load type of connection.

In spite of the seemingly large number of different design layouts these choices allowed, in practice, when one looked at the designs offered at the time, they did not appear to differ greatly from one another, and I will examine the more notable examples of these later in this chapter. Three designs, by McIntosh of the USA, Quad of the UK, and Sonab of Sweden, did, however, strike off on directions of their own, and are worthy of particular consideration.

The McIntosh Amplifier

In the years preceding and following the 1939–1945 war, the McIntosh company of the USA became notable, among the radio and radiogramophone equipment which filled the British radio shops, for producing massive, immaculately engineered equipment offering a superb performance at a high price, for those who wanted and could afford the best. (The company is still in business, and its philosophy appears to be unchanged). At a time when it was thought that amplifier output powers in the range 7–15 watts, and THD figures in the range 0.5–1%, would be entirely adequate for domestic audio purposes, McIntosh offered the circuit shown in Figure 6.1, which had an output power of 50 watts, and a THD figure of 0.2% from 50Hz to 10kHz (McIntosh, F.H., and Gow, G.J., *Audio Engineering*, December 1949).

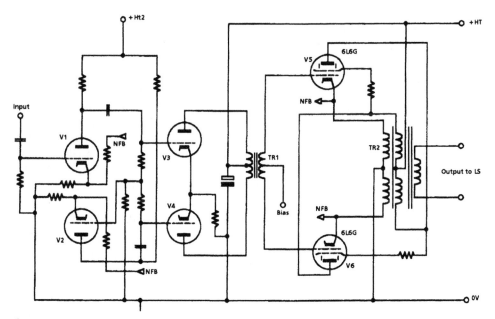

Figure 6.1
The McIntosh system

There were three basic design problems to be solved in this application. The first of these was that obtaining such high output power levels, even with fixed bias, entails driving the output valves in class AB1. This is a condition where the level of negative grid bias is chosen so that one or other of the output valves will be driven into cut-off at some part of the operating cycle. In normal circumstances this will lead to crossover distortion, due, in part, to the imperfect coupling between the two halves of the output transformer primary, and to shifts in the input bias level due to the high output impedance of the driver stage.

McIntosh solved both of these problems by using a trifilar wound output transformer, in which there was very tight inductive coupling between the windings, and the cathodes as well as the anodes of the output valves were coupled to the load. The final problem, that of achieving satisfactory loop stability with two coupling transformers within the loop, was solved in part by taking great care in the design of the coupling transformers to achieve a high winding inductance at the same time as preserving low levels of leakage reactance – the driver transformer is also trifilar wound (i.e. the windings are made by laying the three wires of the separate windings alongside each other and at the same time). The other method used to preserve NFB loop stability was to derive the NFB lines from the cathode of each output valve to that of each input valve rather than from the output to the loudspeaker. The latter arrangement is preferable on technical grounds.

An additional contribution to the close coupling of the two halves of the output stage is provided by connecting the output valve screening grids (G2) to the anodes of the valves on the opposite side of the transformer.

The Williamson Amplifier

The period which saw the introduction of the McIntosh amplifier was one of very great activity on the part of audio amplifier designers, and a number of novel circuit designs were published, of which some were offered specifically for the use of the home constructor. Among these, by far the most celebrated, was the Williamson amplifier, of which the circuit design is shown in Figure 6.2. This was designed by an enthusiastic audiophile who was working for the GEC receiving valve development laboratories at Hammersmith, London. His circuit, first described in a GEC internal memo in 1944, was subsequently published in April/May 1947 (Williamson, D.T.N., *Wireless World*, April 1947, pp. 118–120; idem. May 1947, pp. 161–163).

The springboard for Williamson's design was the prospect of the renewed availability of the GEC KT66 output beam-tetrode for general use. (This valve design had been introduced in 1937, but, in the UK, its use in domestic applications during the 1939–1945 period of hostilities had been restricted due to the overriding priority given to military requirements.) The KT66 beam-tetrode valve was particularly suited to audio output use, in that its output waveform distortion, unlike that of an output pentode in which the distortion was mainly third harmonic, was predominantly second harmonic and, if the output valves were matched, this type of distortion could be substantially reduced by the use of the output valves in push-pull. Moreover, by connecting its G2 to the anode, the KT66 could be made to behave in a manner very closely analogous to that of an output power triode, without any of the problems associated with high power triode construction.

In his introduction to the *Wireless World* articles, Williamson examined the various circuit options available to the audio amplifier designer, and concluded that for the requirements of low harmonic distortion, negligible phase shift within the chosen frequency band (10Hz–20kHz), good transient response and low output resistance to be met, it was essential that the loudspeaker output transformer should be included

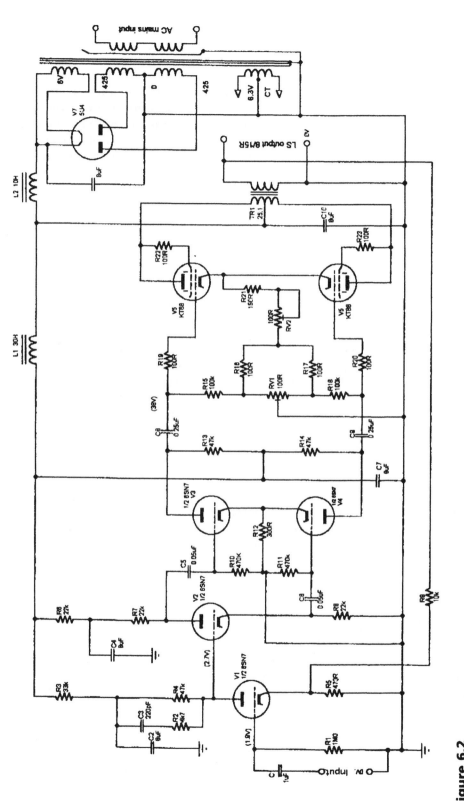

Figure 6.2
The Williamson amplifier (voltages in brackets are AC signal levels)

within the negative feedback loop. This, in turn, required that the transformer primary incremental inductance should be greater than 100H, and the primary:primary and primary:secondary leakage reactances should be less than 33mH. These figures were calculated from the premise that in order to ensure loop stability the phase shift introduced by this component should not exceed 90° at 3.3Hz and 60kHz: a requirement which was easier to meet if the anode impedance of the output valves was not too high, so that the required primary:secondary turns ratio for the output transformer would also not be too large. For a triode connected KT66, R_a would be 2500Ω and the optimum total primary to secondary turns ratio for a 15Ω LS load would be 25:1.

A triode valve, used as a low power voltage amplifier, has a very low level of non-linearity distortion, especially if the output signal voltage is small in relation to the total possible output voltage swing. In Williamson's circuit the output voltage swing at the anode of the input valve, V1, was only 2.7V peak, which is very small in relation to the possible 100V swing at this point.

In order to minimise the LF phase shift within the amplifier, the anode of V1 was directly connected to the split-load phase splitter, V2, and since the DC voltage on V1 anode was about +100V, the voltage on V2 cathode would be about 105V and on its anode would be about +200V. Since V2 is effectively a cathode follower, it will have a signal gain very close to unity and negligible waveform distortion. Moreover, since the same current flows through both R7 and R8, it must follow that if the load impedances are identical, the signal voltages across both of these load resistors would be identical.

Because of the relatively low gain from the input of V1 to the output of the phase splitter, it was necessary to have a further gain stage between this and the output valves, and here, again, a push-pull pair of triodes was used. Since the cancellation of second harmonic distortion requires that the output valves are matched – both in characteristics and in operating conditions – Williamson included preset controls (RV1 and RV2 in Figure 6.2) to allow matching of both grid bias and signal drive levels.

Since the final circuit structure now gave an adequate gain without the use of (electrolytic) cathode bypass capacitors, these were omitted and this provided a small amount of local negative feedback in each gain stage. This helped to improve the gain/bandwidth product and the gain and phase linearity of these stages, and also eliminated a component whose performance was sometimes less than desirable, as well as being prone to long-term deterioration.

Provided that the specification of the output transformer, as given by Williamson, was met, the performance of this amplifier was excellent, and was far ahead of any of its competitors. The articles describing its circuit were reprinted on a world-wide scale, and its performance served as a model to which other designers could aspire. Since it was not easy to improve on this design in terms of performance, most of the competitive designs sought to offer a higher output power than the 15 watts which the

Williamson gave, though, in the UK, many of the commercial designers judged that 12 watts would be adequate for all normal domestic purposes.

Distributed Load Systems

The dilemma which faced audio amplifier designers in their choice of output valves was that of balancing output stage efficiency (and available output power) against harmonic distortion. It was a matter of general observation among the audio enthusiasts of this period that the type of sound quality given by a triode output stage, which mainly generated second harmonic distortion, was preferable to the rather hard or shrill reproduction which was typical of the output pentode, which mainly produced third and fifth harmonics. Output beam-tetrodes, of the KT type, which tended more to a triode distortion characteristic, sounded better than pentodes, but triodes were still the enthusiasts' preference. (With modern beam-tetrodes or pentodes it was possible to get something fairly closely approaching triode characteristics by connecting G2 to the anode.) The problem was in output efficiency, which was only 27% for a push-pull pair of triodes, as compared with nearly 40% for a pair of pentodes using the same HT voltage and anode current.

Understandably, circuit designers began to explore ways in which an output beam-tetrode (or pentode) valve could be made to give a higher output power while still retaining the low distortion characteristic of the triode stage, and it was shown by Hafler and Keroes (*Audio Engineering*, pp. 15–17, November 1951) that this could be achieved by connecting G2 to a tap on the output transformer primary winding, as shown in Figure 6.3. The effect of this method of transformer connection, in terms of relative output power, output impedance and harmonic distortion, is shown graphically in Figure 6.4. Hafler and Keroes termed this the ultralinear (U–L) connection, and this raised the hackles of the linguistic purists, one of whom – I forget who – likened it to the thirteenth chime of a crazy clock, which served only to cast doubt on all that had gone before. Nevertheless the name stuck.

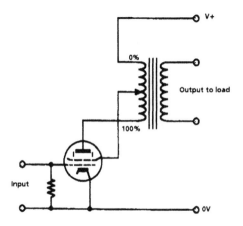

Figure 6.3
The ultralinear connection

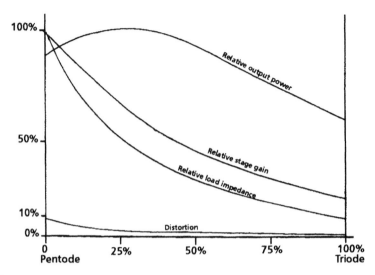

Figure 6.4
Distributed load effects

The practical effect of the U–L connection was to offer the designer a graded choice between the virtues and drawbacks of triode or pentode outputs. In terms of distortion and output (anode) impedance – low values of which made the design of high quality output transformers easier – there was a continual improvement as one moved from pentode to 100% triode, but for maximum output power for a given level of distortion, the 25% tapping was the best. The final point of concern to the designer was that the stage gain in the pentode connected mode was substantially higher than the triode, and this would facilitate the design of the preceding stages of the amplifier. The best compromise value for this was the 40% tap, and this was the ratio favoured for the high quality end of the market.

Much attention was drawn to this technique by the publication by Hafler and Keroes of a circuit which offered ultralinear operation as a means of improving the performance of the Williamson amplifier (*Audio Engineering*, pp. 26–27, June 1952). This prompted an indignant response from Williamson and Walker (the latter being the head of the Acoustical Manufacturing Company, the design of whose Quad audio amplifier used a similar, but allegedly superior, philosophy and had been on sale since 1945). Their response (*Amplifiers and Superlatives*, Williamson, D.T.N., and Walker, P.J., *Audio Engineering*, pp. 75–81, April 1954) gives one of the more comprehensive analyses of this topic, and compares U–L operation with cathode coupled systems of the kind used by both Quad and McIntosh.

The Quad System

This method of operation of the output valve is shown schematically in Figure 6.5. This arrangement offered a number of practical advantages, among which was the fact that the local negative feedback introduced into the cathode circuit, by a winding which was about 10% of that in the anode, reduced the output stage distortion from

2–3%, a typical value for a triode connected output valve, to some 0.7% – a figure which would be reduced still further by overall NFB. It also offered an output stage efficiency which was of the same order as that of an output pentode. Moreover, the local NFB reduced the phase shift due to the output transformer and made the specification for this component less stringent. The full circuit of the Quad 12 watt power amplifier is shown in Figure 6.6. Since this design was offered commercially in 1945, it has the distinction of being one of the first true Hi-Fi power amplifiers, with the NFB loop including the output transformer as well as the whole signal amplifying chain.

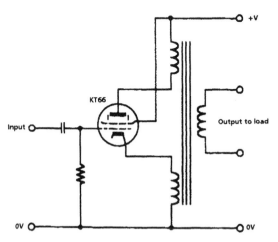

Figure 6.5
The Quad circuit

There are a number of interesting (and surprising) features in the Quad II design, which I have redrawn in Figure 6.6 so that these features are easier to see. Superficially the circuit uses a single pentode amplifying stage followed by a simple phase inverter of the kind shown in Figure 4.5, in which the ratio of R7 to R8 (R4:RV1 in Figure 4.5) is chosen to give the correct output drive voltage at V2 anode. However, there are complications in that the NFB signal, from a tap on the output transformer secondary, is applied to both of the input valve cathode circuits, and while this will appear as negative feedback at V1, it will be positive feedback when applied to V2.

PFB has the effect of increasing both the stage gain and the distortion of the stage to which it is applied, and its use, within the NFB loop, to increase the loop gain and thereby to reduce the overall distortion level has been employed many times.

However, in this instance, the increase in the gain of V2 makes it necessary to reduce the input signal to V2 grid to a much lower level than would normally have been used. The actual values of R7 and R8 would imply a stage gain of 253 for V2, which is rather more than twice that which would have been expected.

Another unusual factor is the cross-coupling, by C1, of the G2 grids of V1 and V2, instead of the more normal separate decoupling of both grids to the ground (0V) or cathode line. This, and the shared cathode bias resistor (R4), helps to restore the equality of the two drive signals applied to the output valves.

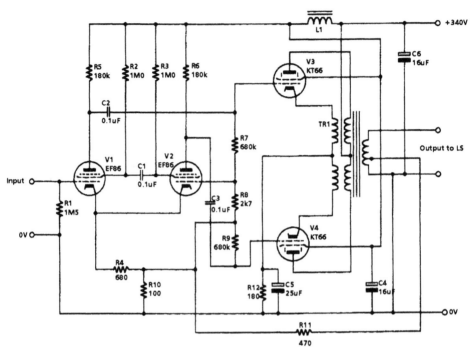

Figure 6.6
The Quad amplifier

The Baxandall Amplifier

There was clearly a degree of rivalry between amplifier designers at this time, since, shortly after Williamson had described his 15 watt design, in 1947, Baxandall proposed in 1948 a circuit design for an amplifier offering a comparable standard of quality at a lower cost – but also over a more limited frequency range (30Hz–15kHz) and with a lower (10 watt) power output – (Baxandall, P.J., *Wireless World*, pp. 2–6, January 1948). There were two innovative features shown in Baxandall's design, of which the first was the decision to use pentode (or tetrode) valves throughout, including a push-pull beam-tetrode output stage. Baxandall argued that the extra gain obtained this way, for the same number of gain stages, would allow more NFB to be applied, without instability, and this would reduce the hum and noise in the system and would also allow a substantial reduction in the overall THD.

The second innovation proposed in this circuit was to derive the overall NFB signal from a separate secondary winding on the output transformer. The absence of any significant output load on this winding greatly reduced the internal phase shifts within this part of the transformer and facilitated the use of quite a high level of NFB – up to 36dB, adjustable by RV3 – without the need for the elaborate and costly design described by Williamson. Unfortunately this idea had been patented by C.G. Mayo, of the BBC Engineering Research Dept., and any commercial manufacturer who sold transformers to Baxandall's design would have to pay royalties to the BBC.

I have shown the circuit of the Baxandall 10W amplifier in Figure 6.7. In terms of its circuit structure, it uses a pair of SP61 high slope pentode valves (V1 and V2) as a floating paraphase phase splitter directly driving an output pair of 6L6s, V3 and V4. These valves were the RCA (USA) equivalent of the Marconi–Osram KT66s and had a somewhat lower permissible anode voltage rating and anode dissipation.

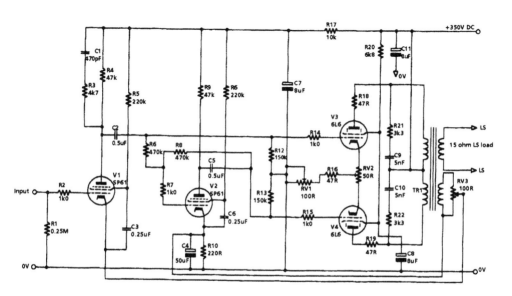

Figure 6.7
The Baxandall amplifier

Interestingly, Baxandall proposed the use of a square wave test signal as a means of uncovering wide-band performance defects, and used this method, with a 50Hz input signal, to show the very considerable improvement which NFB made to the overall LF performance – a possible weak feature in any valve system. It is a pity, in retrospect, that the performance of this amplifier was not tested also with a high frequency square wave – say 10kHz – while driving a reactive load, in that this would have allowed a more direct comparison to be made between this amplifier and contemporary solid state audio designs, which are often tested in this way.*

Baxandall had taken particular care in tailoring the loop phase shifts within this design, and had chosen the values of the coupling capacitors (C2,C5) and those used for screen decoupling (C3,C6) to constrain the amplifier LF phase lead to an

---

* My own measurements, in 1966, on two valve amplifiers, one of commercial origin and one of my own design, showed that the handling of a 10kHz square wave was relatively poor, even on a resistive load. Unfortunately, although the memory remains, the oscilloscope photographs from these tests are no longer available.

adequately low level. A similar function, at HF, was performed by the step network (C1,R3) across the anode load resistor (R4) of V1, and the Zobel networks (R21,C9 and R22,C10) across the two transformer half primaries. The Zobel networks also protected the output transformer (and output valves) against damaging high voltage swings if the amplifier was over-driven, or operated into a very high impedance (or open-circuit) load.

Unfortunately, although the more attractive features of this design were copied, usually in a simplified form, by many amateur constructors, Baxandall's design did not succeed in displacing the Williamson circuit as the leader in public esteem, or in overall sound quality. The weak link, both in terms of the cost and complexity of the output transformer, and in the amplifier performance as a whole, was, of course, the fact that the NFB was drawn from a different winding from that which fed the loudspeaker. This led to the possibility that if the loudspeaker load current was distorted due to the non-uniformities of the moving coil magnetic system – which would usually be the case – the output signal from the amplifier would also be distorted – because of the effective high output impedance of the power amplifier. Baxandall noted this fact in two places in his article describing the design, but blamed the fault on imperfections in the loudspeaker – as if there would ever be a perfect LS unit.

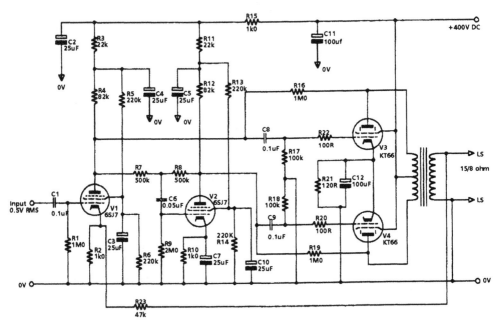

Figure 6.8
20–25W amp (JLH 1951)

However, as I noted, this didn't prevent the general layout of his circuit being copied by various amateur designers. An example of this is given in a circuit of my own, shown in Figure 6.8. This was designed around a relatively inexpensive 20W. output transformer made by the Wharfedale Wireless Works, which had rather poorer figures for primary winding inductance and leakage reactance than Williamson's, but which would nevertheless allow the circuit shown to work quite stably with about 15dB of NFB if local feedback, around the output valves, (R16,R19) restricted their stage gain to about 12˜. I made several of these amplifiers at the time (1951–1952), one of which was made for a local music loving cabinet maker in exchange for a most magnificent oak radiogram cabinet which he then built to my design.

Baxandall's 5 Watt Design

In March 1957, Baxandall returned to the field of audio amplifier design with a simplified, relatively low power (5W) design, because he felt that there was a niche for an inexpensive design intended for use with relatively efficient loudspeaker systems (Baxandall, P.J., *Wireless World*, pp. 108–113, March 1957). In this design, of which I have shown the circuit in Figure 6.9, Baxandall retained both the use of output pentodes and a fairly high level of NFB to keep the residual hum level low, and the total harmonic distortion (THD) below the 0.1% target level at 400Hz and 5 watts output, though he now used the split load phase-splitter system (V2) proposed by Williamson.

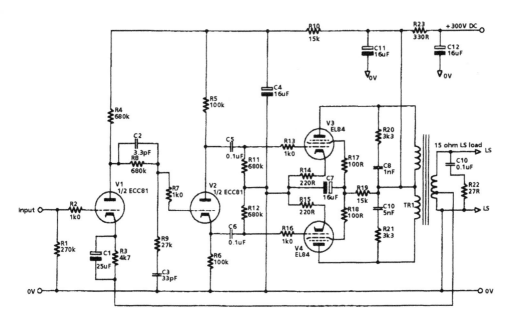

Figure 6.9
Baxandall's simple amp

The use of high levels of NFB with an output transformer of relatively simple construction brought with it the probability of HF instability – which, if it occurred while handling a speech or music signal, would lead to a greater degree of impairment in the audible quality than any possible improvement due to reduced steady-state THD. To provide the very necessary correction to the poor HF loop stability, Baxandall added a further output Zobel network (R22,C10) across the whole secondary of the transformer, in addition to the networks (C8,R20, C10,R21) retained across the output transformer half primaries. Further circuit refinements for the purposes of HF NFB loop compensation were the lag-lead networks (C2,R8 + C3,R9) interposed in the signal line between V1 and V2.

The Radford Valve Amplifiers

These designs were marketed by Radford* Audio Ltd, of Long Ashton, Bristol, over a period of nearly twenty years, and enjoyed a reputation among music lovers which was second to none, mainly because of the very high quality of the output transformer and the general attention to detail in components and design. They were normally available in output power ratings from 7 to 30 watts, though units up to 100 watts could be provided to special order. The circuit I have shown in Figure 6.10 is that of their MA15 15 watt mono version, but the other designs were very similar in form.

In terms of its circuit structure, the output stage uses a push-pull pair of output power pentodes, operating in class A, and coupled to an output transformer which is used in a 40% U–L mode. The output valves are individually cathode biased and are driven by an ECC83 double-triode long-tailed pair phase splitter. Negative feedback is taken from the whole of the output transformer secondary winding to the cathode circuit of V1 via a phase correcting network (R28,C17), of which the component values are switched to suit the output load impedance chosen. (This is also switch selected, but I have omitted the details of this in the interests of clarity.)

Local NFB is applied between anode and grid of V1, and additional loop phase correcting networks are included across V1 anode load (R7,C3) and in the grid drive to the output valves (R16,C8 and R17,C9). Apart from the primary inductance of the transformer the only LF phase error inducing components are C10,R18 and C11,R19, and this ensures loop LF stability.

---

*Arthur Radford, the owner and founder of the firm, was, in his own view, a transformer designer who 'dabbled a bit in electronics'. However, he had not come to terms with 'solid state' electronics, and although his firm manufactured and sold transistor operated audio power amplifiers, they were not of his design, and the amplifier which he preferred, and used in his own home, was the STA25 – a 25 watt stereophonic amplifier which was one of the last of his own creations. I had the great privilege of acquaintanceship with Mr Radford, and was invited to his home on one occasion in the early 1970s so that we could do a comparative listening trial between his own domestic STA25 and my recently designed 75 watt (Hi-Fi News) transistor operated unit. I had expected that the two power amplifiers, which, for the purposes of the trial, were both driven from Radford's own valve preamp, would sound the same, and, so far as I could tell, this was indeed the case. However, I was flattered to find that Mr Radford thought that my design was superior in sound quality to his, and he then invited me, as a consultant, to do some electronic circuit design work for him, which suggested that his judgement was not just a compliment intended to please a guest: not that I thought that such an act would have been in his nature.)

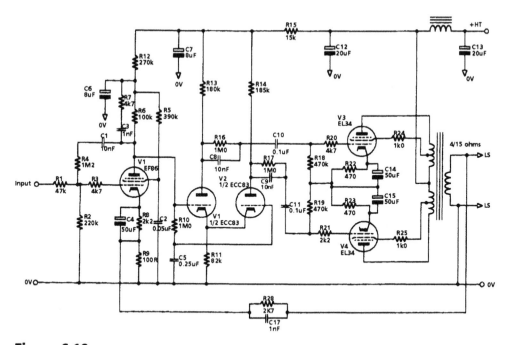

Figure 6.10
The Radford design

The Leak TL/12

This circuit, which I have shown in Figure 6.11, was conservatively rated at 15 watts, and used a circuit which was very similar to that of the Radford, the main differences being in the capacitative – rather than direct – coupling between V1 anode and V2 grid, and that the output valves were triode (rather than U–L) connected. As in the Radford an HF loop phase correction network (R23/C12) was included in the NFB line, and, once again, the values of these components were switched to suit the output transformer secondary ratio chosen. The only other loop compensation component was the 1nF capacitor across R6 in the cathode circuit of V1.

The Mullard 5/10 and 5/20 Designs

The Mullard 5/10, so described because it had an output power of 10 watts and used five valves, including the rectifier, and its companion design with 20 watts output, formed the third of this closely similar group of designs, and, because Mullard Ltd was a wholly owned subsidiary of the Dutch Philips Electronics NV organisation, these circuits could be taken as representing a broad consensus of European audio amplifier design. Unlike the situation in the UK and the USA, beam-tetrode valves were not used, output stages being based solely on pentode valve types.

The design I have shown in Figure 6.12 is the Mullard 5/10, a design which was enormously popular among amateur constructors, and for which output transformers

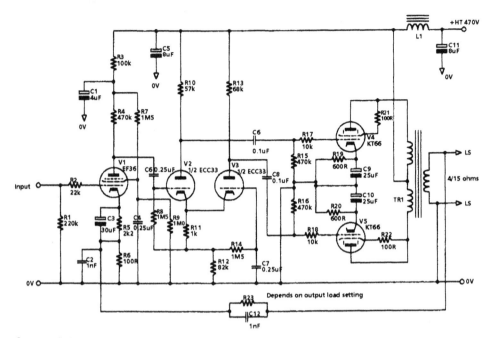

Figure 6.11
The Leak TL12

and mains transformers were available from a wide range of suppliers. The main differences between the 10 and 20 watt units were that the higher powered version used a higher HT line voltage, used EL34s rather than the smaller, pentode connected, EL84s as the output valves, and that the output was U–L connected. Mullard noted that the best performance – presumably in respect of output power – was with the screen tapping points being taken at 20% from the centre, though they did give technical details for the use of the 43% tapping connections, which was an operating mode favoured by the average constructor and transformer supplier.

The quoted harmonic distortion for both these designs, at their rated output powers, was in the range 0.3–0.4%, which, though entirely adequate for most amateur use, did not compete, in the minds of the audio enthusiasts, with the less than 0.1% THD at full output power which was claimed for the Williamson, the Radford or the Leak designs.

Full details of these and other Mullard audio designs were reprinted by Mullard in 1959 in a booklet *Mullard Circuits for Audio Amplifiers* and this was reprinted in 1993 by Audio Amateur Publications, of Peterborough, NH, USA.

The GEC Audio Amplifiers

A major milestone in audio amplifier design was the Williamson amplifier, of 1947, and the British General Electric Company, as the owners of the Marconi/Osram valve manufacturing plant were D.T.N. Williamson's employers. The celebrity which his work had generated, even after Williamson had left the GEC and had joined Ferranti, provided an incentive to their valve development laboratories to develop a range of audio amplifier designs, of which the most significant, from my point of view, was a 30 watt U–L design shown in Figure 6.13, which was basically a somewhat modified and simplified version of Williamson's 15 watt circuit.

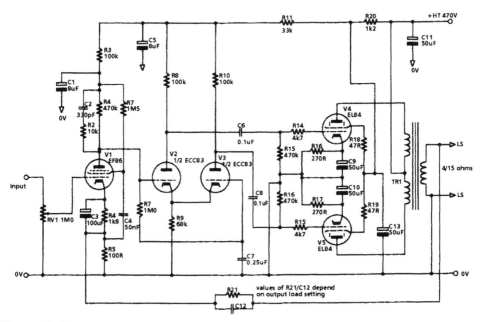

Figure 6.12
The Mullard 5/10 amp

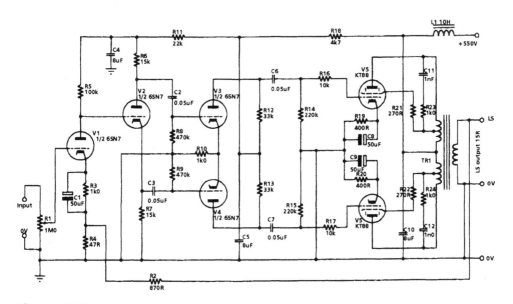

Figure 6.13
GEC 30W Amplifier

This amplifier could be built with KT66s, but gave a better performance, <0.2% THD at 30 watts, if the (then) newly introduced KT88s were used as the output valves. With either KT88s of the recently introduced KT77s (a redesign of the KT66) the power of this amplifier could be increased to 60 watts by employing fixed bias rather than cathode bias. This is done by simply removing the cathode bias resistors of V5 and V6 and returning their cathodes to the 0V rail, and then applying a negative grid voltage of some 48 volts in value to the earthy ends of the grid resistors R14 and R15. In the revised 60 watt circuit the negative grid bias for each output valve was obtained separately from small preset potentiometers connected across a simple 55V negative DC supply.

The GEC 912+ Amplifier

The period from 1950 to 1965 was one of great activity on the part of amateur audio enthusiasts who tried out a wide variety of experimental ideas. The GEC company reflected the mood of the moment and offered a metal cone loudspeaker driver unit, and an economical and easy to build 12–14 watt audio power amplifier – the GEC 912+. The + indicated that it included a simple preamplifier with a gramophone record replay frequency response equalisation stage and a passive tone control circuit. The actual output power depended on the design and efficiency of the output transformer. In the circuit which I have shown in Figure 6.14, I have deleted the input, single valve, preamplifier and tone control circuitry, which would provide the normal input to V1, in the interests of clarity. I have also deleted the presence switch, in the NFB line, with

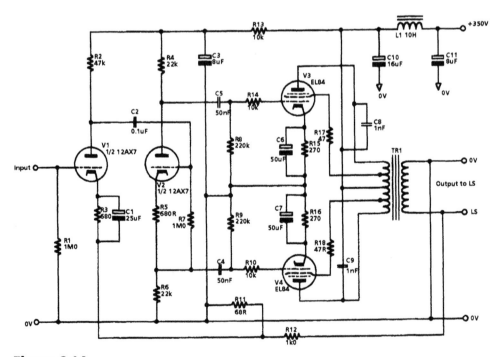

Figure 6.14
The GEC '912+'

which an RC series network could be switched across either R11 or R12, depending on the effect sought.

In the view of the amateur constructor this amplifier was in direct competition with the highly successful Mullard 5/10 design, with the GEC unit offering a somewhat greater power, at a rather worse full output power (typically 0.8% THD) distortion figure. In terms of circuit design, the GEC amplifier used a single triode input amplifier (V1), followed by a split load phase splitter (V2) which provided the drive for the output (U–L connected) pentodes. The total loop feedback was not high, because of the limited gain of the amplifying stages. This was the reason for the relatively poorer THD figure for this design, but the greater loop stability then allowed 'presence control' type tinkering with the frequency response.

Other GEC Audio Designs

It is noteworthy that GEC had produced a number of application reports during the 1950s which covered a wide range of valve amplifier designs in this field, from 3W to 1100W. Most of these designs were gathered together in a booklet, *An approach to audio frequency amplifier design*, originally published in 1957, and reprinted in 1994 by Audio Amateur Publications, of Peterborough, NH, USA. In addition to these design booklets GEC's valve development laboratories continued to produce application reports relating to valve audio amplifier design up to 1979 – by which time almost all the audio design effort had followed the engineering developments in semiconductor technology into the field of the solid state.

It is somewhat to be regretted that amongst all this effort little novel circuitry emerged from the GEC's audio engineers, except in the very high power designs, and these were aimed at public address use, rather than high quality domestic audio applications, and mainly were not of a type which would allow the use of overall loop NFB to attain low levels of harmonic and intermodulation distortion, and mainly didn't employ overall NFB.

The Brimar 25P1 Design

This is, in many ways, similar to the Leak TL12 or the Radford STA 25 designs, but has some interesting design features, and achieves a very good performance with a relatively simple circuit layout. The quoted performance is 25 watts output at less than 0.1% THD, at 1kHz, a bandwidth of 25Hz–20kHz, ±0.2dB, an input sensitivity of 480mV for 25W output, and a hum and noise figure of >85dB referred to 20 watts output. Other figures which are very good for a valve amplifier design are: Rise time 6μS, Phase shift, <20° at 25kHz, and Z_{out} <0.4Ω from 50Hz to 10kHz.

Structurally, the circuit consists of a high gain pentode input stage, with an HF phase compensation network, C2,R24, across its anode load resistor. This drives a pair of triodes, V2,V3, which act as a floating paraphase phase splitter, and which in turn drive an output push-pull pair of beam–tetrodes, V4,V5. (The output valves specified were STC type 5B255Ms, but these were closely similar to the contemporary KT66s,

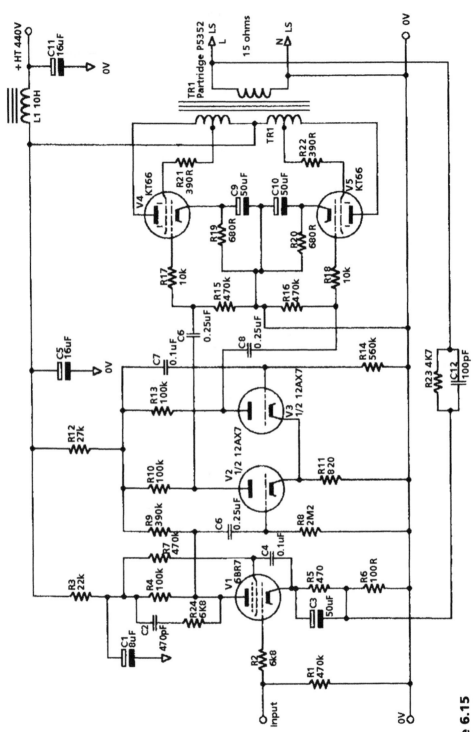

Figure 6.15
The Brimar 25P1 amp

which I have shown.) An innovative feature is the use of a resistor, R12, which is common to both V2 and V3 anode circuits to derive the paraphase input signal to V3. This leads to rather less loss of gain than the rather more common grid-derived paraphase input signal. A further subsidiary NFB loop, which contributes a further −10dB to the total, and assists in achieving a good loop stability, is drawn from V2 anode circuit by way of R9. The circuit shown is for a 15Ω output load, and the values of R23 and C12 are those appropriate for this load figure.

I consider this circuit to be an excellent example of valve audio design practice, along with the Brimar SP55 preamplifier quoted in Chapter 5, which could well serve as models to those designers wishing to return to valve operated audio systems

The Sonab Design

Sadly, from the point of view of the historian, there have been relatively few genuinely original valve circuit designs in the audio amplifier field, but the OA6 Mk 1 design offered by Sonab, of Sweden, during the period 1966–1970, is a remarkable example of circuit innovation. I have shown the basic layout of this design in Figure 6.16, simplified by omitting the various NFB loops as well as the power supplies to the second grids of the output pentodes. A somewhat unusual feature of the design is that it requires a conducting path between its LS output terminals, and it will not work, electrically, without such a low resistance output load. I have assumed that this will always be the case for the purposes of the circuit analysis.

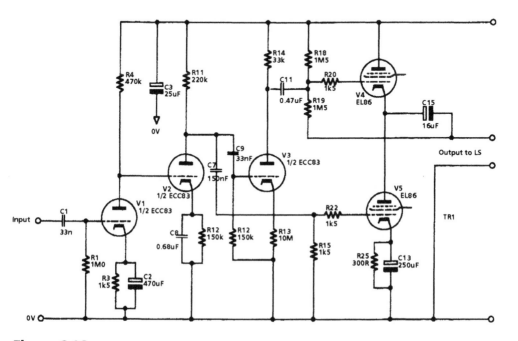

Figure 6.16
The Sonab system

The output stage is connected as a White type cathode follower (see Jones, M., *Valve Amplifiers*, Newnes, 1995, p. 91). In this layout, the output cathode follower, V4, feeds an active load, V5, which is driven in phase opposition to V4. This phase inversion of the signal is provided by an additional triode amplifier stage, V3, inserted between V2 and V4. The cathode potential of V4 is arranged to sit at about half the input HT line voltage by taking its grid to a voltage divider network, R18/R19, which is connected between V+ and the 0V line via the LS load. Because of the low impedance of this output stage – approximately half that of V4 acting as a cathode follower on its own – it is possible, with this layout, to drive a high impedance LS unit (say, 50–100 ohms), or a pair of headphones, directly, without the need for an impedance matching output transformer.

In the complete circuit of the OA6/1 design, shown in Figure 6.16, several further innovative features are added. Firstly, the input valve, V1, is operated in a virtual earth configuration, with the input and NFB signals being summed at its grid. This shunt type NFB is theoretically superior to the more common series type NFB, where the feedback signal is returned to the input valve cathode, because it is free from any possible common-mode errors. Secondly, both voltage mode NFB (taken across the LS load, via NFB1) and current mode NFB (derived, via NFB2, from the current flowing through the LS load by a current transformer, TR1) signals are employed. In addition, a third NFB path is taken directly from the amplifier output to the cathode of V2.

Since Sonab felt that the availability of suitable high impedance LS units might be limited, they provided, as the other half of the OA6 design, a mirror image circuit in which only voltage feedback was used, and in order to provide a low Z output the amplifier output was coupled to the load via a conventional output transformer. Since the voltage NFB signal in the low Z half of this amplifier was taken from the primary of the output transformer, this would leave the design open to the inherent fault Baxandall found, in his design of Figure 6.7, that the inevitable loudspeaker load non-linearities would distort the waveform at the amplifier output – regardless of the absence of distortion in the circuit of the amplifier itself.

By modern standards, the resistance values employed in this design, such as in the input and NFB paths to V1, are very high, and would lead to a relatively high background thermal noise level, as well as some DC level instability, due to valve aging processes, as a consequence of high value grid leak resistances.

Increasing Available Output Power

As has been seen, the use of the U–L output transformer connection provides a simple means for increasing the output power obtainable from a given design, but, with suitable protection arrangements to prevent inadvertent operation of the system without output grid bias, fixed bias operation can also substantially increase the available output. For example, in the case of U–L connected KT77, at 430V HT, an output power of 34 watts is quoted at 2.5% THD, while a similar fixed bias operation will allow an output of 48 watts at 1% THD. For the KT88, the comparable figures, at 460V HT, are 50 watts and 80 watts respectively.

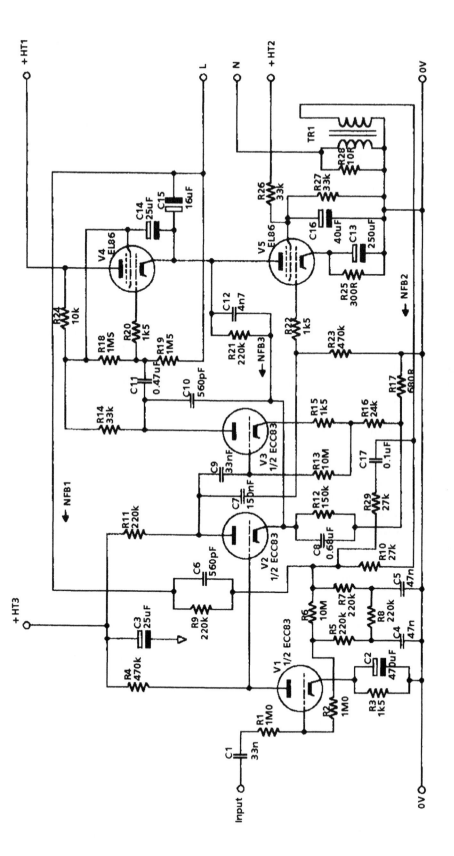

Figure 6.17
The Sonab OQ6/1 amp

Conclusions

With hindsight it is easy to see the pitfalls into which some of the pioneering valve amplifier designers fell. The need, for example, for the amplifier LS output to be taken from the winding, and the whole of that winding, from which the loop NFB signal was drawn, was not generally appreciated until the late 1950s, this allowed the situation in which the amplifier could be virtually distortion free, until the LS load was connected. By the time of the Radford and Mullard designs it was accepted that the NFB signal must be taken from the LS output connections, and if the transformer primary:secondary ratio was changed to suit different LS loads then the attenuator network (cf. R28 and C17 in Figure 6.10) in the feedback path must also be changed to suit this new condition.

In a valve amplifier, the output transformer is a source of waveform distortion, so, for an output THD of better than 0.1% over the bulk of the audible frequency spectrum, overall loop NFB, including the output transformer, must be employed. The amount of this which it was needful to employ would depend mainly on the quality of the output transformer, and on the type of output valve used. For a well made LS transformer, and triode connected output valves, this degree of linearity could be met with some 20dB of NFB. For U–L connected beam-tetrodes, this performance would require some 26dB of NFB, while pentode or beam-tetrode output valves, an NFB ratio of nearer 30dB was needed. Beam tetrodes were, in general, slightly more linear than pentodes.

In order to have an adequate gain margin to allow sensible amounts of NFB to be employed – without the need for very high input signal voltages – either a driver stage must be used between the phase splitter and the output valves, or a phase splitter with both a high gain and a large permissible output voltage swing, such as a floating paraphase or a long-tailed pair, is required.

Unless the required gain can be obtained with relatively few gain stages, the use of NFB will lead to problems with HF and LF instability. Poor loop stability at the LF end of the spectrum will probably cause motor-boating which will be audible. HF instability may cause continuous oscillation at frequencies which are outside the hearing range, which will spoil the sound quality and may damage the amplifier or the LS units. An insidious problem with poor loop stability margins is that the instability may be sporadic, triggered by certain combinations of signal frequency, output signal level and load reactance. This is most easily checked by the use of an oscilloscope, a variable frequency square wave input and a range of non-resistive loads. A number of HF/LF phase compensation networks have been shown in Figures 6.7, 6.9 and 6.10.

CHAPTER 7

SOLID STATE VOLTAGE AMPLIFIERS

The solid state device technologies which are available to the amplifier designer fall, broadly, into three categories: bipolar junction transistors (BJTs), and junction diodes; junction field effect transistors (FETs); and insulated gate FETs – usually referred to as MOSFETs (metal oxide silicon FETs), because of their method of construction. These devices are available both in P type – operating from a negative supply line – and N type – operating from a positive line. BJTs and MOSFETs are also available in small-signal and larger power versions, while FETs and MOSFETs are manufactured in both enhancement-mode and depletion-mode forms. Predictably, this allows the contemporary circuit designer very considerable scope for circuit innovation, by comparison with electronic engineers of the past, for whom there was only a very limited range of vacuum tube devices.

In addition, there is a very wide range of integrated circuits (ICs) which are complete functional modules in some (usually quite small) individual packages. These are designed both for general purpose use, such as operational amplifiers, and for more specific applications, such as voltage regulator devices, current mirrors, current sources, phase sensitive rectifiers, and an enormous variety of designs for digital applications, which mostly lie outside the scope of this book.

In the case of discrete devices, I think it is unnecessary for the purposes of audio amplifier design to understand the physical mechanisms by which the devices work provided that their would-be user has a reasonable grasp of their operating characteristics and limitations, and, above all, a knowledge of just what is available.

Junction Transistors

These are nearly always three-layer devices, fabricated by the multiple and simultaneous vapour phase diffusion and etching of small and intricate patterns on a large, thin slice of very high purity single crystal silicon. A few devices are still made in germanium, mainly for replacement purposes, and some VHF components are made in gallium arsenide, but these will not, in general, lie within the scope of this book. The fabrication techniques may be based on the use of a completely undoped (intrinsic) slice of silicon, into which carefully controlled quantities of impurities are diffused through an appropriate mask pattern from both sides of the slice. These will be described in the manufacturers' literature as double-diffused or triple-diffused and so on.

In a later technique, evolved by the Fairchild Instrument Corporation, all the diffusions were made from one side of the slice. These devices were called planar and had, normally, a better HF response and more precisely controlled characteristics than, for example, equivalent double-diffused devices. In a further, more recent, technique, also due to Fairchild, the silicon slice will have been made to grow a surface layer of uniformly doped silicon on the exposed side (which will usually form the base region of a transistor) and a single diffusion was then made into this doped layer, to form the emitter junction. This technique was called epitaxial and led to transistors with superior characteristics, especially at HF. Since this is the cheapest BJT fabrication process it will normally be used wherever it is practicable, and if no process is specified it may reasonably be supposed to be a planar-epitaxial type.

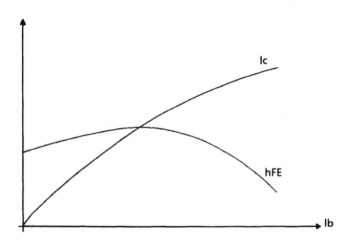

Figure 7.1
BJT non-linearity

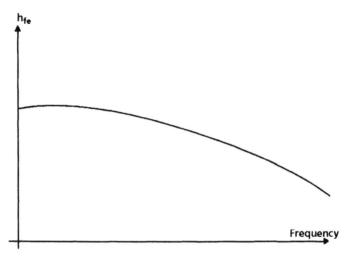

Figure 7.2
Decrease in h_{fe} with frequency

In contrast to a thermionic valve, which is a voltage controlled device, the BJT is a current operated one. So while a change in the base voltage will result in a change in the collector current this has a very non-linear relationship to the applied base voltage. In comparison to this, the collector current changes with the input current to the base in a relatively linear manner. Unfortunately, this linear relationship between I_c and I_b tends to deteriorate at higher base current levels, as shown in Figure 7.1. This relationship between base and collector currents is called the current gain, and for AC operation is given the term h_{fe}, and its non-linearity is an obvious source of distortion when the device is used as an amplifier. Alternatively, one could regard this lack of linearity as a change of h_{FE} (this term is used to define the DC or LF characteristics of the device) as the base current is changed. A further problem of a similar kind is the change in h_{fe} as a function of signal frequency, as shown in Figure 7.2.

However, as a current amplifier (which generally implies operation from a high impedance signal source) the behaviour of a BJT is vastly more linear than when used as a voltage amplifying stage, for which the input voltage/output current relationships are shown for an NPN silicon transistor as line 'a' in Figure 7.3. (I have included, as line 'b', for reference, the comparable characteristics for a germanium junction transistor, though this would normally be a PNP device with a negative base voltage and a negative collector voltage supply line.) By comparison with, say, a triode valve, whose anode current/grid voltage relationships are also shown as line 'c' in Figure 7.3, the BJT is a grossly non-linear amplifying device, even if some input (positive in the case of an NPN device) DC bias voltage has been chosen so that the transistor operates on a part of the curve away from the non-conducting initial region.

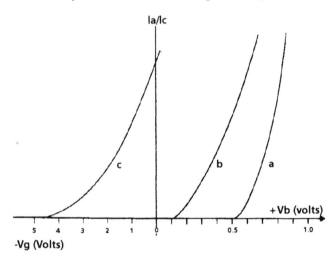

Figure 7.3
Comparative characteristics of valve, germanium and silicon based BJTs

Control of Operating Bias

There are three basic ways of providing a DC quiescent voltage bias to a BJT, which I have shown in Figure 7.4. In the first of these methods, shown in Figure 7.4a, an arrangement which is fortunately seldom used, the method adopted is simply to connect an input resistor, R1, between the base of the transistor and some suitable voltage source. This voltage can then be adjusted so that the collector current of the transistor is of the right order to place the collector potential near its desired operating voltage. The snag with this scheme is that transistors vary quite a lot from one to another of nominally the same type, so this would require to be set anew for each individual device. Also, if the operating temperature changes, the current gain of the device (which is temperature sensitive) will alter – and, with it, the collector current of Q1 and its working potential. The arrangement shown in Figure 7.4b is somewhat preferable in that a high current gain transistor, or one working at a higher temperature, will pass more current, and this will lower the collector voltage of Q1,

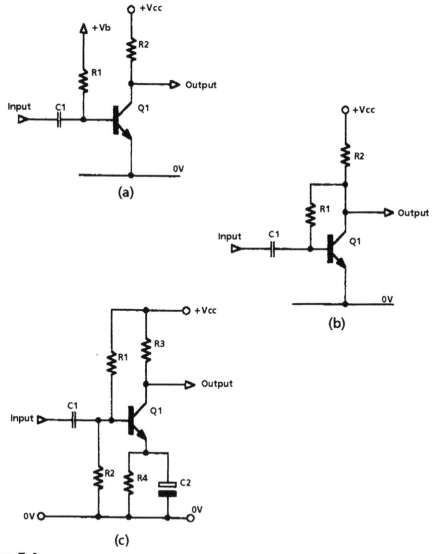

Figure 7.4
Biasing circuits

which will, in turn, reduce the bias current flowing through R1. However, this also provides NFB and will limit the stage gain to a value somewhat less than $R1/Z_{in}$.

The method almost invariably used in competently designed circuitry is that shown in Figure 7.4c, or some equivalent layout. In this, a potential divider (R1,R2) having an output impedance which is low in relation to the base impedance of Q1, is used to provide a fixed DC base potential. Since the emitter will, by emitter-follower action, sit at a potential, depending on emitter current, which is about 0.6V below that of the base, the value of R4 will then determine the emitter and collector currents, and the operating conditions so provided will hold good for almost any broadly similar device used in this position. Since the emitter resistor would cause a significant reduction in stage gain, as seen in the equivalent analysis of valve cathode bias systems, it is customary to bypass this resistor with a capacitor, C2, which is chosen to have an impedance which is low in relation to R4 and R3.

Stage Gain

The stage gain of a BJT, used as a simple amplifier, can be determined from the relationship:

$$\frac{V_{out}}{V_{in}} = \frac{h_{fe}R_L}{R_s + r_i}$$

where R_s is the source resistance, R_L is the collector load resistor, h_{fe} is the small-signal (AC) current gain, and r_i is the internal emitter-base resistance of the transistor. An alternative, and somewhat simpler, approach is similar to that which one would use for a pentode valve gain stage, in which

$$V_{out}/V_{in} = g_m R_L$$

where the g_m of a typical modern planar epitaxial silicon transistor will be in the range 25–40mS per mA of collector current. Because the g_m of the junction transistor is so high, high stage gains can be obtained with a relatively low value of load resistor. For example, a small-signal transistor with a supply voltage of 15V, a 4k7 collector load resistor and a collector current of 2mA will have a low frequency stage gain, for a relatively low source resistance, of some 300⁻. If some way can be found for increasing the load impedance, without also increasing the voltage drop across the load, very high gains indeed can be achieved – up to 2500 with a junction FET acting as a high impedance constant current load (see Linsley Hood, J., *Wireless World*, September 1971, pp. 437–441).

A predictable, but interesting aspect of stage gain is that the higher the gain which can be obtained from a circuit module, the lower the distortion in this which will be due to the input device. This is so because if increasingly small segments are taken from any curve they will progressively approach more closely to a straight line in their form. This allows a very low THD figure, much less than 0.01% at 2V rms output, over the frequency range 10Hz–20kHz, to be obtained from the simple NPN/PNP feedback pair shown in Figure 7.5, which would have an open loop gain of several thousand. The distortion contributed by Q2 will be relatively low because of the high effective source resistance seen by Q2 base. A similar low level of distortion is given by the amplifier layout (bipolar transistor with constant current load) described above, because of the very high stage gain of the amplifying transistor and the consequent utilisation of only a very small portion of its Ic/Vb curve.

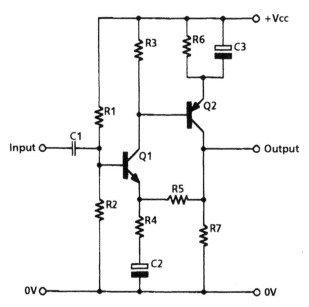

Figure 7.5
NPN/PNP feedback pair

Basic Junction Transistor Circuit Configurations

As in the case of the thermionic valve, there are a number of layouts, in addition to the simple single transistor amplifier shown in Figure 7.4, or the two stage amplifier of Figure 7.5, which can be used to provide a voltage gain or to perform an impedance transformation function. There is, for example, the grounded base layout of Figure 7.6, which has a very low input impedance, a high output impedance, and a very good HF response. This circuit is far from being only of academic interest in the audio field, in that it can provide, for example, a very effective low input impedance amplifier circuit for a moving coil pick-up cartridge. I showed a circuit of this type, dating from about 1980, in an earlier book (*Audio Electronics*, Newnes, 1995, p. 133).

The cascode layout, described in Chapter 4, is also very widely used as a voltage amplifier stage, using a circuit arrangement of the kind shown in Figure 7.7a. As in the case of the valve amplifier stage, this circuit gives very good input/output isolation and an excellent HF performance due to its freedom from capacitative feedback from output to input. It can also be rearranged, as shown in Figure 7.7b, so that the input stage acts as an emitter-follower, which gives a very high input impedance.

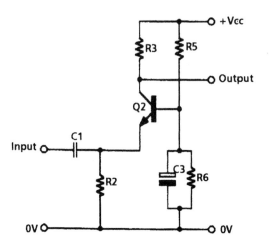

Figure 7.6
Grounded base stage

The long-tailed pair layout, shown in its simplest form in Figure 7.8a, gives a very good input/output isolation, and also, since it is of its nature a push-pull layout, it gives a measure of reduction in even-order harmonic distortion. Its principle advantage, and the reason why this layout is normally used, is that it allows, if the tail resistor (R1) is returned to a –ve supply rail, both of the input signal ports to be referenced to the 0V line – a feature which is enormously valuable in DC amplifying systems. The designer may sometimes seek to improve the performance of the circuit block by the use of a high impedance (active) tail, in place of a simple resistor, as shown in Figure 7.8b. This will lessen the likelihood of unwanted signal breakthrough from the –ve supply

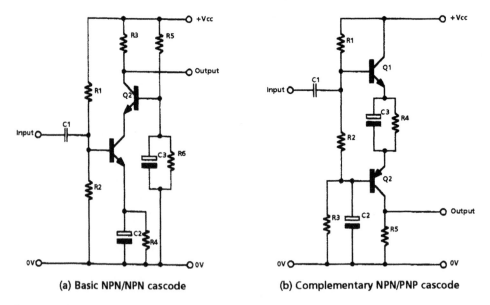

(a) Basic NPN/NPN cascode   (b) Complementary NPN/PNP cascode

Figure 7.7
Cascode layouts

rail, as well as ensuring a greater degree of dynamic balance, and signal transfer, between the two halves.

Although like all solid state amplifying systems it is free from the bugbears of hum and noise intrusion from the heater supply of a valve amplifying stage – likely in any valve amplifier where there is a high impedance between cathode and ground – it is less good from the point of view of thermal noise than a similar single stage amplifier, partly because there is an additional device in the signal line, and partly because the gain of a long-tailed pair layout will only be half that of a comparable single device gain stage. This arises because if a voltage increment is applied to the base of Q1, then Q1 emitter will only rise half of that amount due to the constraint from Q2, which will also see, but in opposite phase and halved in size, the same voltage increment. This allows, as in the case of the valve phase splitter, a very close similarity, but in opposite phase, of the output currents at Q1 and Q2 collectors.

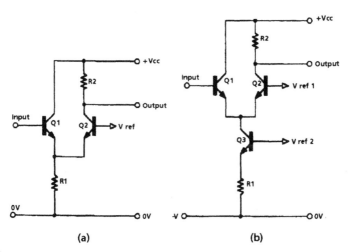

Figure 7.8
Long-tailed pair layouts

Emitter-follower Systems

These are the solid state equivalent of the valve cathode follower layout, though offering superior performance and greater versatility. In the simple circuit shown in Figure 7.9 (the case shown is for an NPN transistor, but a virtually identical circuit, but with negative supply rails, could be made with a similar PNP transistor) the emitter will sit at a quiescent potential about 0.6V more negative than that of the base, and this will follow, quite accurately, any signal voltage excursions applied to the base. (There are some caveats in respect of capacitive loads, and I will explore these potential problems under the heading of slew rate limiting.) The output impedance of this circuit is low, because this is approximately equal to $1/g_m$, and the g_m of a typical small-signal, silicon BJT is of the order of 35mA/V (35mS) per mA of emitter current. So, if Q1 is operated at 5mA, the expected output impedance, at low frequencies, will be $1/(5\,\tilde{}\,35)$ kilohms, or 5.7 ohms; a value which is sufficiently smaller than any likely

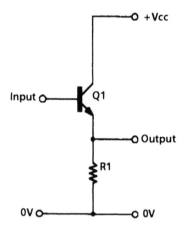

Figure 7.9
Emitter-follower

value for R1 that the presence of this resistor will not greatly affect the output impedance of the circuit.

The output impedance of a simple emitter-follower can be still further reduced by the circuit elaboration shown in Figure 7.10, known as a compound emitter-follower. In this, the output impedance is lowered in proportion to the effective current gain of Q2, in that, by analogy with the output impedance of an operational amplifier with overall NFB, any change in the potential of Q1 emitter, brought about by an externally impressed voltage, will result in an opposing change in the collector current of Q2. This layout gives a comparable result to that of the Darlington pair, of two transistors, in cascade, connected as emitter-followers, shown in Figure 7.11, except that the arrangement of Figure 7.10 will only have an input/output DC offset equivalent to a

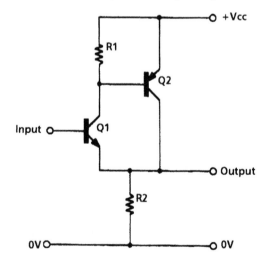

Figure 7.10
Compound emitter-follower

single emitter-base forward voltage drop, whereas the layout of Figure 7.11 will have two, giving a combined quiescent voltage offset of the order of 1.3–1.5V. Nevertheless, in commercial terms, the popularity of power transistors, internally connected as a Darlington pair, mainly for use in the output stages of audio amplifiers, is great enough for a range of single chip Darlington devices to be offered by the semiconductor manufacturers.

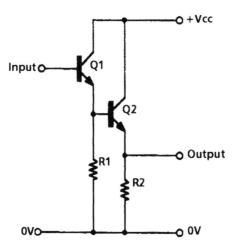

Figure 7.11
Darlington pair

Thermal Dissipation Limits

Unlike a thermionic valve, the active area of a BJT is very small, in the range 0.5mm≈ for a small signal device to 4mm≈ or more for a power transistor. Since the physical area of the component is so small – this is a quite deliberate choice on the part of the manufacturer because it reduces the individual component cost by allowing a very large number of components to be fabricated on a single monocrystalline silicon slice – the slice thickness must also be kept as small as possible – values of 0.15–0.5mm are typical – in order to assist the conduction of any heat evolved by the transistor action away from the collector junction to the metallic header on which the device is mounted.

Whereas in a valve, in which the internal electrode structure is quite massive, and heat is lost by a combination of radiation and convection, the problem of overheating is usually the unwanted release of gases trapped in its internal metalwork, in a bipolar junction transistor the problem is the phenomenon known as thermal runaway. This can happen because the potential barrier of a P-N junction (that voltage which must be exceeded before current will flow in the forward direction) is temperature dependent, and decreases with temperature. Since there will be unavoidable non-uniformities in the doping levels across the junction this will lead to non-uniform current flow through the junction sandwich, with the greatest flow taking place through the hottest region. If the ability of the device to conduct heat away from the junction is inadequate to prevent the junction temperature rising above permissible levels, this process can become cumulative. This will result in the total current flow through the device being funnelled though some very small area of the junction, and this may permanently damage the transistor. This malfunction is termed secondary breakdown, and the operating limits imposed by the need to avoid this failure mechanism are shown in Figure 7.12. Field effect devices do not suffer from this type of failure.

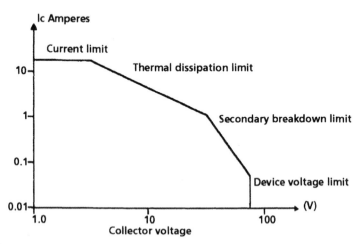

Figure 7.12
Bipolar breakdown limits

Junction Field Effect Transistors (JFETs)

These are, almost invariably, depletion mode devices – which means that there will be some drain current at a zero applied gate-source potential. This current will decrease in a fairly linear manner as the reverse gate-source potential is increased, giving an operating characteristic which is, in the case of an N-channel JFET, very similar to that of a triode valve, as shown in curve 'c' of Figure 7.3. Like a thermionic valve, the operation of the device is limited to the range between drain (or anode) current cut-off and gate (or grid) current. In the case of the JFET, this is because the gate-channel junction is effectively a silicon junction diode – normally operated under reverse bias conditions. If the gate source voltage exceeds some 0.6V in the forward direction, it will conduct, which will prevent gate voltage control of the channel current.

P-channel JFETs are also made, though in a more limited range of types, and these have what is virtually a mirror image of the characteristics of their N-channel equivalents, though in this case the gate-source forward conduction voltage will be of the order of –0.6V, and drain current cut-off will occur in the gate voltage range of +3 to +8V. Although Sony did introduce a range of junction FETs for power applications, these are no longer available, and typical contemporary JFETs cover the voltage range (maximum) from 15 to 50V, mainly limited by the gate-drain reverse breakdown potential, and with permitted dissipations in the range 250–400mW. Typical values of g_m (usually called g_{fs} in the case of JFETs) fall in the range of 2–6mS.

JFETs mainly have good high frequency characteristics, particularly the N-channel types, of which there are some designs capable of use up to 500MHz. Modern types can also offer very low noise characteristics, though their very high input impedance will lead to high values of thermal noise if their input circuitry is also of high impedance; however, this is within the control of the circuit designer. The internal noise resistance of a JFET, R(n), is related to the g_{fs} of the device by the equation:

$$R(n) \text{ ohms} \approx 0.67/g_{fs}$$

and the value of g_{fs} can be made higher by paralleling a number of channels within the chip. The Hitachi 2SK389 dual matched-pair JFET achieves a g_{fs} value of 20mS by this technique, with an equivalent channel thermal noise resistance of 33 ohms (Linsley Hood, J., Low noise systems, *Electronics Today International*, July 1992, pp. 42–46).

Although JFETs will work in most of the circuit layouts shown for junction transistors, the most significant difference in the circuit structure is due to their different biasing needs. In the case of a depletion mode device it is possible to use a simple source bias arrangement, similar to the cathode bias used with an indirectly heated valve, of the kind shown in Figure 7.13. As before, the source resistor, R3, will need to be bypassed with a capacitor, C2, if the loss of stage gain due to local NFB is to be avoided. As with a pentode valve, which the junction FET greatly resembles in its operational characteristics, the simplest way of calculating stage gain is by the relationship:

$$A \approx g_{fs} R_L$$

The device manufacturers will frequently modify the structure of the JFET to linearise its V_g/I_d characteristics, but, in an ideal device, these will have a square-law relationship, as defined by:

$$g_{fs} = I_d/V_g \approx I_{dss}[1 - (V_{gs}/V_{gc})] \approx /V_g$$

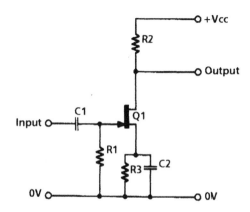

Figure 7.13
Simple JFET biasing system

For a typical JFET operating at 2mA drain current, the g_{fs} value will be of the order of 1–4mS, which would give a stage gain of up to 40 if R2, in Figure 7.13, is 10kΩ. This is very much lower than would be given by a bipolar junction transistor, and is the main reason why they are not often used as voltage amplifying devices in audio systems, unless their very high input impedance (typical values are of the order of $10^{12}\Omega$) or their high, and largely constant, drain impedance characteristics are advantageous.

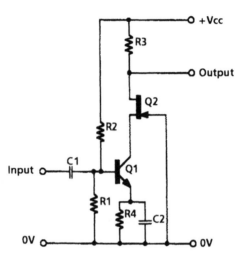

Figure 7.14
Bipolar/FET cascode

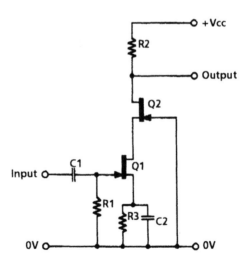

Figure 7.15
FET/FET cascode

The real value of the JFET emerges in its use with other devices, such as the bipolar/FET cascode shown in Figure 7.14 or the FET/FET cascode layout of Figure 7.15. In the first of these, the use of the JFET in the cascode connection confers the very high output impedance of the JFET, and the high degree of output/input isolation characteristic of the cascode layout, coupled with the high stage gain of the BJT. The source potential of the JFET (Q2) will be determined by the reverse bias appearing across the source/gate junction, and could typically be of the order of 2–5V, and this will define the collector potential applied to Q1. A further common application of this type of layout is that in which the cascode FET (Q2 in Figure 7.15) is replaced by a high voltage BJT. The purpose of this arrangement is to allow a JFET amplifier stage to operate at a much higher rail voltage than would be allowable to the FET on its own, and this layout is often found as the input stage of high quality audio amps.

A feature which is very characteristic of the JFET is that for drain potentials above about 3V the drain current, for a given gate voltage, is almost independent of the drain voltage, as shown in Figure 7.16. Bipolar junction transistors have a high

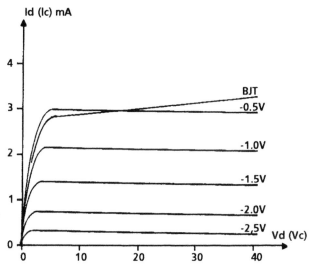

Figure 7.16
Drain current characteristics of junction FET

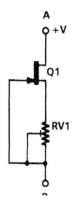

Figure 7.17
Current source layout

characteristic collector impedance, but their I_c/V_c curve for a fixed base voltage, also shown, for comparison, in Figure 7.16, is not as flat as that of the JFET. The very high dynamic impedance of the JFET resulting from this very flat I_d/V_d relationship encourages the use of these devices as constant current sources, of the kind shown in Figure 7.17. In this form the JFET can be treated as a true two-terminal device, from which the output current can be adjusted, with a suitable JFET, over the range of several milliamperes down to a few microamperes, by means of RV1.

Insulated Gate FETs (MOSFETs)

These devices, usually called MOSFETs, are by far the most widely available, and most widely used, of all the field effect transistors. They normally have a rather worse noise figure than an equivalent JFET, but, on the credit side, they have rather more closely controlled operating characteristics. The range of types available covers the very small-signal, low working voltage components used for VHF amplification in TVs and FM tuners (for which applications a depletion-mode dual-gate device has been introduced which has very similar characteristics to those of an RF pentode valve) to high power, high working voltage devices for use in the output stages of audio amplifiers, as well as many other high power and industrial applications. They are made both in depletion- and enhancement-mode forms (the former having gate characteristics similar to that of the JFET, while the latter description refers to the style of device in which there is normally no drain current in the absence of any forward gate bias), in N-channel and P-channel versions, and, at the present time, in voltage and dissipation ratings of up to 1000V and 600W respectively.

All MOSFETs operate in the same manner, in which a conducting electrode (the gate) situated in proximity to an undoped layer of very high purity single-crystal silicon (the channel), but separated from it by a very thin insulating layer, is caused to induce an electrostatic charge in the channel, which will take the form of a layer of mobile electrons or holes. In small-signal devices this channel is formed on the surface of the chip between two relatively heavily doped regions, which will act, respectively, as the source and the drain of the FET, while the conducting electrode will act as the current controlling gate.

Although modern photolithographic techniques are capable of generating exceedingly precise diffusion patterns, the length of the channel formed by surface masking techniques in a lateral MOSFET will be too long to allow a low channel 'on' resistance. For high current applications the semiconductor manufacturers have therefore evolved a range of vertical MOSFETs. In these, very short channel lengths are achieved by sequential diffusion processes from the surface, which are then followed by etching a V or U shaped trough inwards from the surface so that the active channel is formed across the exposed edge of a thin diffused region. Since this channel is short in length its resistance will be low, and since the manufacturers generally adopt device structures which allow a multiplicity of channels to be connected electrically in parallel, channel 'on' resistances as low as 0.008Ω have been achieved.

Like a JFET, the MOSFET would, left to itself, have a square-law relationship between gate voltage and drain current. However, in practice this is affected by the device geometry, and many modern devices have a quite linear I_d/V_g characteristic, as shown in Figure 7.18 for an IRF520 power MOSFET.

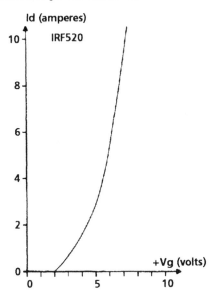

Figure 7.18
Power MOSFET

The basic problem with the MOSFET is that of gate/channel over-voltage breakdown – in which the thin insulating layer, of silicon oxide or silicon nitride, between the gate electrode and the channel breaks down. If this happens the gate voltage will no longer control the drain current, and the device is defunct. Because it is theoretically possible for an inadvertent electrostatic charge, such as might arise in respect to the ground if a user were to wear nylon or polyester fabric clothing and well-insulated shoes, it is common practice in the case of small-signal MOSFETs for protective diodes to be formed on the chip at the time of manufacture. These could be either Zener diodes or simple junction diodes connected between the gate and the source or the source and drain as shown in Figure 7.19.

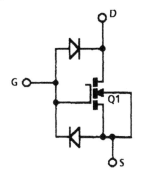

Figure 7.19
Diode gate protection

In power MOSFETs these protective devices are seldom incorporated into the chip. There are two reasons for this: firstly that the effective gate/channel area is so large that the associated capacitance is high, and this would then require a relatively large inadvertently applied static charge to generate a destructive gate/channel voltage (typically >40V), and secondly that such protective diodes could, if they were triggered into conduction, cause the MOSFET to act as a four-layer thyristor, and become an effective electrical short-circuit. However, there are usually no performance penalties which will be incurred by connecting some external protective Zener diode in the circuit to prevent the gate/source or gate/drain voltage exceeding some safe value, and this is a common feature in the output stages of audio power amplifiers using MOSFETs.

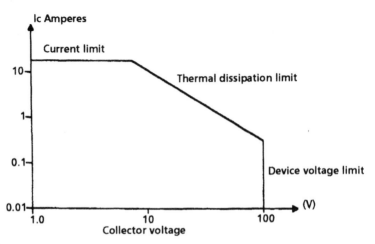

Figure 7.20
Power MOSFET SOAR limits

Apart from the possibility of gate breakdown, which, in power MOSFETs, always occurs at less than the maker's quoted voltage except at zero drain current, MOSFETs are quite robust devices and the safe operating area rating (SOAR) curve of these devices, shown for a typical MOSFET in Figure 7.20, is free from the threat of secondary breakdown whose limits are shown, for a power BJT, in Figure 7.12. The reason for this freedom from localised thermal breakdown in the MOSFET is that the mobility of the electrons (or holes) in the channel decreases as the temperature increases, and this gives all FETs a positive temperature coefficient of channel resistance.

Although it is possible to propose a mathematical relationship between gate voltage and drain current, as was done in the case of the JFET, with MOSFETs the manufacturers tend to manipulate the diffusion pattern and construction of the device to linearise its operation, which leads to the type of performance (quoted for an actual device) shown in Figure 7.21. However, as a general rule, the g_{fs} of a MOSFET will increase with drain current, and a forward transconductance (slope) of 10S/A is quoted for an IRF140 at an I_D value of 15 amperes.

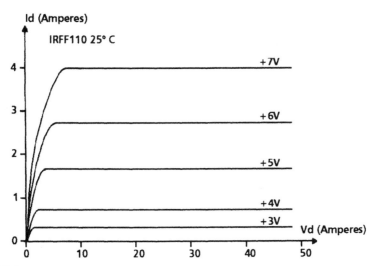

Figure 7.21
MOSFET characteristics

Power BJTs vs. Power MOSFETs as Amplifier Output Devices

Some rivalry appears to have arisen between audio amplifier designers over the relative merits of bipolar junction power transistors, as compared with power MOSFETs. Predictably, this is a mixture of advantages and drawbacks. Because of the much more elaborate construction of the MOSFET, in which a multiplicity of parallel connected conducting channels is fabricated to reduce the conducting 'on' resistance, the chip size is larger and the device is several times more expensive both to produce and to buy. The excellent HF characteristics of the MOSFET, especially the N-channel V and U MOS types, can lead to unexpected forms of VHF instability – which can, in the hands of an unwary amplifier designer, lead to the rapid destruction of the output devices. On the other hand, this excellent HF performance, when properly handled, makes it much easier to design power amplifiers with good gain and phase margins in the feedback loop, where overall NFB is employed. By contrast, the relatively sluggish and complex characteristics of the junction power transistor can lead to difficulties in the design of feedback amplifiers with good stability margins.

Also, as has been noted, the power MOSFET is intrinsically free from the problem of secondary breakdown, and an amplifier based on these does not need the protective circuitry which is essential in amplifiers with BJT output devices if failure is to be avoided when they are used at high power levels with very low impedance or reactive loads. The problem here is that the protective circuitry may cut in during high frequency signal level peaks during the normal use of the amplifier and this can lead to audible clipping. (Incidentally, the proponents of thermionic valve based audio amplifiers have claimed that the superior audible qualities of these, by comparison with transistor based designs, are due to the absence of any overload protection circuitry which could cause premature clipping, and to their generally more graceful behaviour under sporadic overload conditions).

A further benefit enjoyed by the MOSFET is that it is a majority carrier device, which means that it is free from the hole-storage effects which can impair the performance of power junction transistors and make them sluggish in their turn-off characteristics at high collector current levels. However, on the debit side, the slope of the V_g/I_d curve of the MOSFET is less steep than that of the V_b/I_c curve of the BJT and this means that the output impedance of power MOSFETs used as source followers is higher than that of an equivalent power BJT used as an emitter follower. Other things being equal, a greater amount of overall negative feedback (i.e. a higher loop gain) must therefore be used to achieve the same low amplifier output impedance with a power MOSFET design than would be needed with a power BJT one. If a pair of push-pull output source/emitter followers is to be used in a class AB output stage, more forward bias will be needed with the MOSFET than with the BJT to achieve the optimum level of quiescent operational current, and the discontinuity in the push-pull transfer characteristic will be larger in size, though likely to introduce, in the amplifier output signal, lower rather than higher order crossover harmonics.

U and D MOSFETs

I have, so far, lumped all power MOSFETs together in considering their performance. However, there are, in practice, two different and distinct categories of these, based on their constructional form, and these are illustrated in Figure 7.22. In the V or U MOS devices – these are just different names for what is essentially the same system, depending on the profile of the etched slot – the current flow, when the gate layer has been made sufficiently positive (in the case of an N-channel device) to induce a mobile electron layer, will be essentially vertical in direction, whereas in the D-MOS or T-MOS construction the current flow is T shaped from the source metallisation pads across the exposed face of the very lightly doped P region to the vertical N–/N+ drain sink. Because it is easier to manufacture a very thin diffused layer (= short channel) in the vertical sense than to control the lateral diffusion width, in the case of a T-MOS device, by surface masking, the U-MOS devices are usually much faster in response than the T-MOS versions, but the T-MOS equivalents are more rugged and more readily available in complementary (N-channel/P-channel) forms.

All power MOSFETs have a high input capacitance, typically in the range 500–2500pF, and since devices with a lower conducting resistance ($R_{ds/on}$) will have achieved this quality because of the connection of a large number of channels in parallel, each of which will contribute its own element of capacitance, it is understandable that these low channel resistance types will have a larger input capacitance. Also, in general, P-channel devices will have a somewhat larger input capacitance than an N-channel one. The drain/gate capacitance – a factor which is very important if the MOSFET is used as a voltage amplifier – is usually in the range 50–250pF. The turn-on and turn-off times are about the same (in the range 30–100nS) for both N-channel and P-channel types, mainly determined by the ease of applying or removing a charge from the gate electrode. If gate-stopper resistors are used – helpful in avoiding UHF parasitic oscillation, and avoiding latch-up in audio amplifier output source-followers – these will form a simple low-pass filter in conjunction with the device input capacitance, and will slow down the operation of the MOSFET.

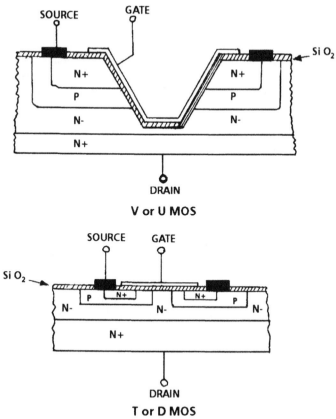

Figure 7.22
MOSFET design styles

Although circuit designers tend to be rather lazy about using the proper symbols for the components in the designs they have drawn, enhancement-mode and depletion-mode MOSFETs should be differentiated in their symbol layout, as shown in Figure 7.23. As a personal idiosyncracy I also prefer to invert the symbol for P-channel field-effect devices, as shown, to make this polarity distinction more obvious.

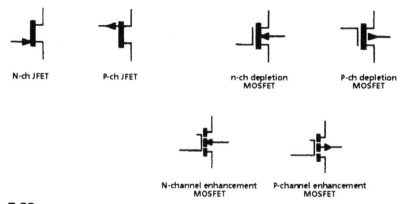

Figure 7.23
MOSFET symbols

Useful Circuit Components

By comparison with the situation which existed at the time when most of the pioneering work was done on valve operated audio amplifiers, the design of solid-state amplifier systems is greatly facilitated by the existence of a number of circuit artifices, contrived with solid state components, which perform useful functions in the design. I have shown a selection of the more common ones below.

Constant current (CC) sources

A simple two-terminal CC source is shown in Figure 7.17, and devices of this kind are made as single ICs with specified output currents. By comparison with the discrete JFET/resistor version, the IC will usually have a higher dynamic impedance and a rather higher maximum working voltage. In power amplifier circuits it is more common to use discrete component CC sources based on BJTs, since these are generally less expensive than JFETs and will be available higher working voltages. The most obvious of these layouts is that shown in Figure 7.24a, in which the transistor, Q1, is fed with a fixed base voltage – in this case derived from a Zener or avalanche diode, though any suitable voltage source will serve – and the current through Q1 is constrained by the value chosen for R1, in that if it grows too large the base-emitter voltage of Q1 will diminish, and Q1 output current will fall. Designers seeking economy of components will frequently operate several current source transistors and their associated emitter resistors (as Q1/R1) from the same reference voltage source.

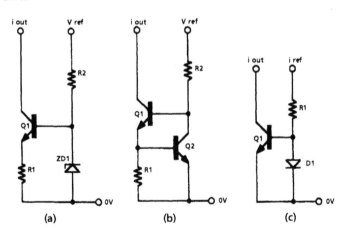

(a)   (b)   (c)

Figure 7.24
Constant current sources

In the somewhat preferable layout shown in Figure 7.24b, a second transistor, Q2, is used to monitor the voltage developed across R1, due to the current through Q1, and when this exceeds the base emitter turn-on potential (about 0.6V) Q2 will conduct and will steal the base current to Q1 provided from V_{ref} through R2. In the very simple layout shown in Figure 7.24c, advantage is taken of the fact that the forward potential of a P-N junction diode, for any given junction temperature, will depend on the current flow through it. This means that, if the base emitter area and doping characteristics of Q1 are the same as those for the P-N junction in D1 (which would, obviously, be easy to arrange in the manufacture of ICs) then the current (i_{out}) through Q1 will be caused to mimic that flowing through R1, which I have labelled i_{ref}. This particular action is called a current mirror, and I have shown several further versions of these in Figure 7.25.

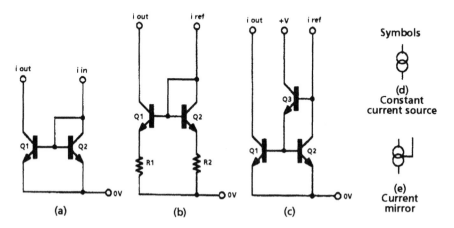

Figure 7.25
Current mirror circuits

Current mirror (CM) layouts

This allows, for example, the output currents from a long-tailed pair to be combined, which increases the gain from this circuit. In the version shown in Figure 7.25a, two matched transistors are connected with their bases in parallel, so that the current flow through Q2 will generate a base-emitter voltage drop which will be precisely that which is needed to cause Q1 to pass the same current. If any doubt exists about the similarity of the characteristics of the two transistors, as might reasonably be the case for randomly chosen devices, equality of the two currents can be assisted by the inclusion of equal value emitter resistors (R1,R2) as shown in Figure 7.25b. For the perfectionist, an improved three-transistor current mirror layout is shown in Figure 7.25c. Commonly used circuit symbols for these devices are shown in Figures 7.25d and 7.25e.

Circuit Oddments

Several circuit modules have found their way into amplifier circuit design, and I have shown some of the more common of these in Figure 7.26. Both the DC bootstrap, shown as 7.26a, and the JFET active load, shown as 7.26b, act to increase the dynamic impedance of R1, though the DC bootstrap – which can, of course, be constructed using complementary devices – has the advantage of offering a low output impedance. The amplified diode, shown as Figure 7.26c, is a device which is much used as a means of generating the forward bias required for the transistors used in a push-pull

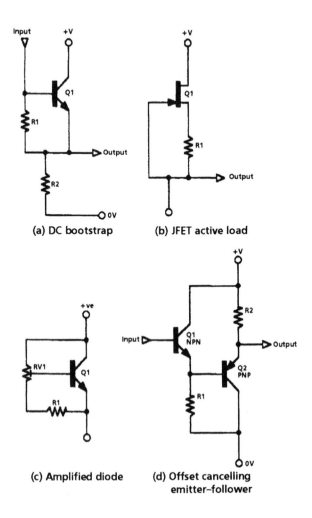

(a) DC bootstrap

(b) JFET active load

(c) Amplified diode

(d) Offset cancelling emitter–follower

Figure 7.26
Circuit oddments

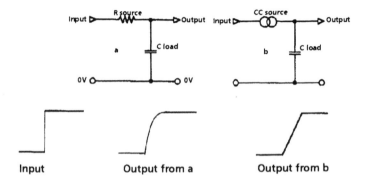

Input

Output from a

Output from b

Figure 7.27
Cause of slew rate limiting

pair of output emitter followers, particularly if it is arranged so that Q1 can sense the junction temperature of the output transistors. It can also be used, over a range of relatively low voltages, as an adjustable voltage source, to complement the fixed voltage references provided by Zener and avalanche diodes, band-gap references (IC stabilisers designed to provide extremely stable low voltage sources) and the wide range of voltage stabiliser ICs. Finally, when some form of impedance transformation is required, without the V_{be} offset of an emitter follower, this can be contrived as shown by putting two complementary emitter followers in series. This layout will also provide a measure of temperature compensation.

Slew Rate Limiting

This is a potential problem which can occur in any voltage amplifier or other signal handling stage in which an element of load capacitance (which could simply be circuit stray capacitance) is associated with a drive circuit whose output current has a finite limit. The effect of this is shown in Figure 7.27. If an input step waveform is applied to the network 'a', then the output signal will have a waveform of the kind shown at 'a', and the slope of the curve will reflect the potential difference which exists, an any given moment, between the input and the output. Any other signal which is present at the same time will pass through this network, from input to output, and only the high frequency components will be attenuated.

On the other hand, if the drive current is limited, the output waveform from circuit 7.27b will be as shown at 'b' and the slope of the output ramp will be determined only by the current limit imposed by the source and the value of the load capacitance. This means that any other signal component which is present, at the time the circuit is driven into slew rate limiting, will be lost. This effect is noticeable, if it occurs, in any high quality audio system, and gives rise to a somewhat blurred sound – a defect which can be lessened or removed if the causes (such as too low a level of operating current for some amplifying stage) are remedied. It is prudent, therefore, for the amplifier designer to establish the possible voltage slew rates for the various stages in any new design, and then to ensure that the amplifier does not receive any input signal which requires rates of change greater than the level which can be handled. A simple input integrating network of the kind shown in Figure 7.27a will often suffice.

CHAPTER 8

EARLY SOLID STATE AUDIO AMPLIFIER DESIGNS

Towards the end of the 1950s, transistors ceased to be merely an interesting laboratory curiosity, and emerged into the component catalogues as valid and usable devices, for a number of low power and low voltage applications. However, few people seriously considered them as possible competitors for valves in higher power applications, or possible options for use in high quality audio systems. There were a number of good reasons for this, of which the major one was that transistors, at that time, mostly meant germanium PNP diffused junction devices, and these were very temperature sensitive, both in respect of the output (collector) current for a given forward bias, and in respect of the electrical characteristics of the device itself. This latter problem arose because, if the diffused junction regions within the transistor got sufficiently hot, say >180°C, the diffusion processes, by which they were manufactured, would continue, and the internal junction boundaries would shift.

At this time, a germanium transistor would be made by taking a small, thin wafer of single crystal germanium containing trace quantities of arsenic or antimony, as an N-type impurity, and then heating the wafer so that two small pellets of indium – previously spot welded facing each other on opposite sides of the wafer – would cause P-type impurity diffusion zones to move in towards each other from opposite sides of the wafer, as shown in Figure 8.1, to form a PNP device. Since the high frequency response of the device, its current gain and its breakdown voltage would depend on the thickness of the zone (the base junction) separating these two regions, if this changed during use it would cause the circuit performance to change also. (In principle, an NPN transistor could also be made by the same technique, and would, presumably, suffer from the same snags. However, neither arsenic nor antimony – the only practicable N-type electron donor materials – were found to behave very satisfactorily as alternatives to indium.)

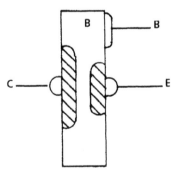

Figure 8.1
Diffused junction transistor

Because of their small size and low supply voltage requirements, germanium PNP devices offered, within the constraints imposed by their sensitivity to heat, a number of advantages for use in things like hearing aids and portable radios. For reasons of economy in power consumption, and the need to avoid much heat dissipation in the output transistors, a typical circuit for a low power audio amplifier – having, say, an output power of 500mW – would use a circuit of the kind shown in Figure 8.2, in which the output devices were operated in class AB, with a very low quiescent collector current. Inevitably, this arrangement led to a fairly high level of crossover distortion, and because the circuit used both driver and output transformers, it would be difficult to employ a useful amount of NFB to reduce its distortion, but perhaps, as the output stage for a portable radio, its performance would be thought to be adequate.

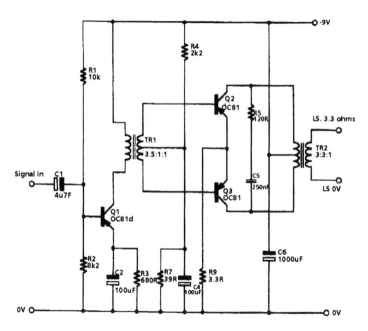

Figure 8.2
Transformer coupled amp

However, the circuit of Figure 8.2 is a very simple one, and much better designs are possible, as, for example, in the 15 watt power amplifier shown in Figure 8.3, due to Mullard Ltd (*Reference Manual of Transistor Circuits*, 2nd edition, 1961, pp. 178–180). In this, by rearranging the circuit somewhat, it has proved possible to dispense with the output LS coupling transformer entirely, so that if the driver transformer is reasonably well designed, it will be possible to employ a small measure of overall NFB, from LS output to signal input, to reduce the distortion and make the frequency response of the circuit more uniform. The quoted performance for this design was: THD <4%, bandwidth 150Hz–7kHz, ±1.5dB. The way that NFB is applied to the power amplifier, as shown, makes the overall gain dependent on the ratio of R7 to the external circuit impedance seen at Q1 base.

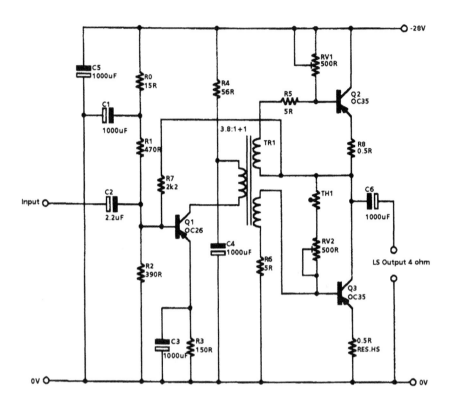

Figure 8.3
15 watt audio amplifier

The Lin Circuit

As has been seen in the case of the valve audio amplifiers described above, the output transformer is a bulky and expensive component whose performance is likely to have a crucial effect on the overall performance of the amplifier design. Since the output impedance of a transistor, used as an emitter-follower, could well be less than an ohm, it should be possible to implement the basic audio amplifier structure shown in Figure 4.1 (a variable gain voltage amplifier stage driving the loudspeaker via some means of impedance conversion) by the use of a pair of push-pull emitter-followers, rather than a transformer, as the output impedance-matching mechanism.

However, there was at this time an additional requirement – that these output emitter-followers should be based on output transistors of the same polarity. This need arose because although a few NPN transistors were available these were all small-signal types, and were quite unsuitable for use as one half of an output emitter-follower. All of these problems appeared to have been solved at a stroke by the ingenious circuit layout proposed by Lin (Lin, H.C., *Electronics*, pp. 173–175, September 1956), and is shown in Figure 8.4. In this, the output emitter-followers were what Lin termed a quasi-complementary pair, in which the upper half (Q2,Q4) was a conventional Darlington pair and the lower half (Q3,Q5) was a compound emitter-follower. This allowed the output voltage, at C4, to follow the amplified signal voltage at the collector of Q1, but at a low enough impedance to drive a 16 ohm LS directly.

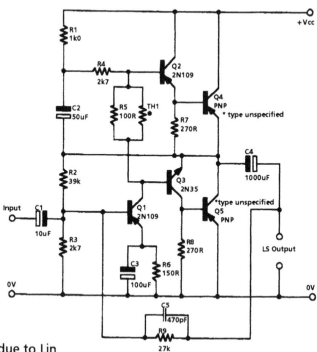

Figure 8.4
Audio amplifier due to Lin

The performance of Lin's design (THD <1% at 400Hz at 6W output, bandwidth of 30Hz–15kHz, ±1.5dB), while not yet as good as could be obtained from a run of the mill valve audio amplifier, nevertheless offered a workable design option. As would be expected, the stability of the voltages and currents in this design was not very good, though Lin had employed DC NFB, via R2, in addition to the AC NFB through R9 and C5, to hold the collector voltage of Q1 to a suitable value. The forward bias applied to the output quasi-complementary emitter-followers, which would need to be about 0.2V, at 25°C, was provided by the voltage drop across R5, caused by the collector current of Q1, and an attempt was made to compensate for changes in the junction temperature of the output transistors by connecting a negative temperature coefficient thermistor (TH1) across R5.

As can be seen from Figure 8.5, germanium junction transistors have a less abrupt turn-on characteristic than silicon ones, and since the output devices will probably run warm, their actual V_b/I_c graph is more likely to be that of the dashed line curve than the solid line, 20°C, one – and this will act to lessen the magnitude of the potential crossover-type discontinuities which lurk within any push-pull system.

However, the somewhat unpredictable performance of germanium transistors of that time, together with their proneness to thermal runaway, discouraged audio amplifier manufacturers from making commercial designs of this type. This had to wait for another five years for the introduction of silicon transistors made by variants of the

Fairchild planar process. These became available initially in small signal versions, and, because the manufacturing techniques favoured this, in NPN (+ve rail) constructions. These allowed the design of high quality, low noise, low distortion, small–signal gain stages which needed no setting-up adjustments, and which, in my view at the time, were a great improvement, in terms of freedom from mains hum and microphony, on their thermionic valve predecessors.

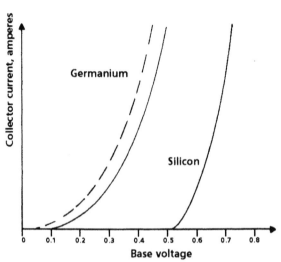

Figure 8.5
Turn-on characteristics

The implementation of audio power amplifier designs with predictable and stable performance characteristics demanded equally reliable and robust output transistors, which meant, in practice, those made using silicon planar construction, and when these became available, they were offered principally as NPN types, so it was using these in Lin-type quasi-complementary circuit arrangements that the first high fidelity solid-state audio amplifiers were made. Unfortunately, this approach led to a type of malfunction which was overlooked by the designers at the time, but which fairly soon became the subject of hostile comment from the users of this new Hi-Fi equipment, and this was the problem of output stage asymmetry.

Quasi-complementary Output Stage Asymmetry

This problem is illustrated in Figure 8.6. The input base voltage vs. collector current relationship of a simple NPN/NPN Darlington pair based on silicon junction transistors is shown in the upper right-hand quadrant of the drawing, and the equivalent characteristics of a silicon transistor compound emitter-follower are shown in the lower left-hand quadrant. Not only are these curves different in slope – which makes the push-pull transfer characteristics asymmetrical, even when operated at the optimum quiescent bias, as shown in the diagram 'a' – but, since the output impedance of an emitter-follower is approximately $1/g_m$, and the slope of the curve in the lower

LH quadrant is markedly steeper than that of the upper RH, the output impedance will be different as well; a factor which will become apparent on low output impedance loads, such as, for example, LS driver units at some parts of their frequency response curve.

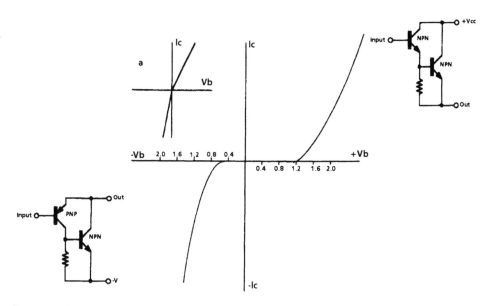

Figure 8.6
Output pair asymmetry

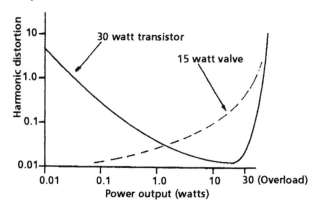

Figure 8.7
Comparative THD curves for transistor and valve amplifiers

Inevitably, this asymmetry in the transfer curves of the two halves of the output stage leads to a degree of residual crossover distortion, which is worsened if the chosen quiescent current setting (the choice of which will always be a matter of some compromise, because what would be the best setting for one half of the output pair would not be the best for the other) is not the optimum value. Since the circuit layout of the amplifier is likely to be fairly simple, with a limited number of phase-shifting elements, it is possible to use a large amount of NFB to reduce the measured full output power total harmonic distortion (THD) level. Since it is inconvenient, on a production assembly line, to have to adjust the output quiescent currents of each half of a stereo amplifier, some designers – having noted that, in an amplifier using a high degree of NFB, the actual quiescent current setting which was chosen made relatively little difference to the measured value of the residual full output power THD – opted either to use some fixed value which was rather less than the optimum, or to use no forward bias at all. This could lead to the situation shown in Figure 8.7, in which the THD could be quite low at the rated power output, but would worsen as the output power level was reduced.

I have shown a typical solid state audio amplifier design of this period, the middle 1960s, in Figure 8.8. The circuit I have chosen is that of the Leak Stereo/Delta 70. This is a typical example of design thinking of that time, and most of the contemporary Hi-Fi shops would have similar designs from a wide range of audio amplifier manufacturers on their shelves.

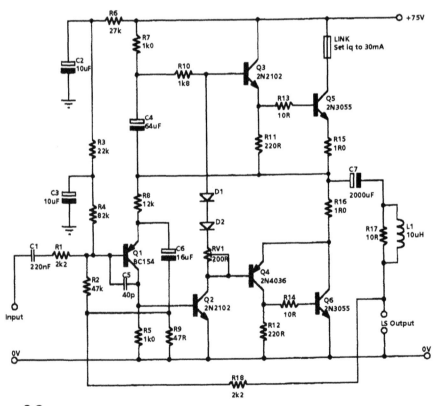

Figure 8.8
Leak Delta 70 amplifier

Listener Fatigue

It was fairly generally accepted that there was a difference in the sound quality given by the new transistor amplifiers and that of the valve amplifiers that they sought to supplant, and this quickly led to the emergence of two camps in the Hi-Fi field, those who liked the new sound, and those who rejected it, and described the tonal quality as hard, or thin or clinical. On the other side were those who argued that, since the output power available was now greater, and the full output power bandwidth and distortion figures were both better than those of earlier systems, what the listener was now hearing was the actuality of the music, and not some rounded-off version, all of whose rough edges had been removed by the inadequacies of the amplifier output transformer. The anti solid state protagonists retaliated by describing the new technology as giving 'transistor sound' and complaining that it caused listener fatigue.

In reality, the new solid state amplifiers suffered from a number of shortcomings, which were largely overlooked by the design engineers because they occurred in areas which had not previously been regions of concern. The first of these, already mentioned, was that of the dependence of the distortion level on the output power. In valve designs operating in class A (that condition in which the amplifying devices are conducting for the whole of the signal voltage excursion) it could be taken for granted that the worst output THD figure would occur just below the onset of clipping, and this would decrease, ultimately disappearing into the noise background, as the output power decreased. However, as had been seen in Figure 8.7, a solid state amplifier operating under zero forward bias (class B) conditions would have a distortion figure which would worsen as the output power was reduced.

An additional factor is also shown in Figure 8.7, in respect of the output power available before the onset of signal peak clipping. A valve amplifier which used only a modest amount of NFB, say 15dB or less, would have a distortion figure which would worsen only gradually as it was driven into overload, and if the listener was prepared to accept a moderate level of peak clipping, the valve amplifier could actually sound louder than the apparently higher powered transistor version. The relatively soft clipping of the traditional valve amplifier, when driven into overload, is one of the more highly valued characteristics of this system, in the view of some contemporary users. This is mainly a feature of the low level of NFB which is used in some such designs, coupled with their freedom from latch up, a frequent feature of badly designed solid state amplifiers.

The use of large amounts of NFB to reduce the apparent distortion level of the amplifier, especially under class B operation, leads to two further problems, of which the first is that when the output transistors are cut off, the system gain is zero and consequently the amount of NFB applied through the feedback loop is also zero. In practice, this means that if the signal voltage swing takes it through the zero voltage axis it will pass into a dead zone, beyond which the full amplifier gain will operate to urge the voltage swing across to the opposite conduction region. However, in the dead zone the amplifier is switched off, and any low level signals which are present in this region will be lost – thus justifying the allegations of the thinness of tone of the amplifier, a characteristic feature of zero-biased or under-biased output stage operation.

In the circuit design of Figure 8.8, the maker's recommended quiescent current setting is 30mA, although, in practice, experience would suggest that the optimum current for the upper Darlington pair will be of the order of 80–100mA, and the lower compound emitter-follower stage would be optimally biased at an I_q of some 40–50mA. Using a higher bias current setting than the optimum would, as shown in diagram of Figure 8.9, lead to a worsening of the THD above some low power level, determined by the actual I_q setting, in exchange for a substantial improvement in the distortion at very low output power levels as the output stages effectively returned to class A operation. Such an amplifier performance might look worse on the specification sheet, but could be more pleasant to listen to.

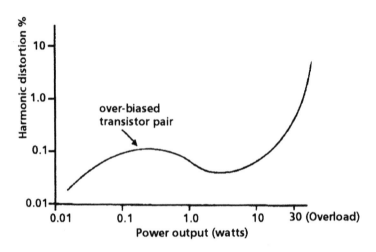

Figure 8.9
Effect on THD of increased value of quiescent current

The second incipient problem in solid state amplifier designs of this period was that of inadequate stability margins in the feedback loop. Like most of the other problems, this was worsened by the lack of symmetry of the output stage, in that not only were the dynamic (amplitude and rate of change related) electrical characteristics of the transistor itself frequency, temperature and current dependent, but they would also vary depending on which of the two output emitter-follower groups was conducting. To add to this complex mix of difficulties, the LS load which was coupled to the amplifier had a reactance which was continuously variable, dependent on the frequency and amplitude levels of the input signal. All of this presented a substantially greater challenge to the loop stability of the amplifier than that offered by the conventional resistive dummy load – so that amplifiers which behaved quite stably on the test bench might well pass through regions of instability under live conditions with an LS load, which would lead to the occurrence of brief bursts of HF oscillation buried, but not hidden from the ears of the listener, within the signal.

Design engineers working in this field in the mid-1960s were acutely aware of the need for some improvement on the type of performance given by the standard quasi-complementary (Q/C) output pair, and a number of options were explored with this aim in mind.

Alternative Circuit Choices

In the absence of PNP power transistors, or, when such devices had become available, obtainable only in relatively low voltage and low power versions, there were two possible options open to the designer – to improve the symmetry of the quasi-complementary pair circuit, so that NPN output power transistors could be used exclusively without audible performance penalties, or to bias the output emitter-followers so that they operated in class A, in which condition the crossover distortion would be greatly reduced. The first of these routes was chosen by the Acoustical Manufacturing Company in their Quad 303 power amplifier, in which they elaborated the two-transistor quasi-complementary pair into a triplet, as shown in Figure 8.10a (report, *Wireless World*, April 1968, p. 67). In this layout, an almost exact symmetry of the V_{in}/I_{out} curves was obtained, though there was still some small difference between the two triples in respect of the optimum quiescent current, of which the mean value was only, in any case, about 4mA. This low value of optimum I_q led to the minor drawback that it did not allow any significant margin of operation in class A, which would act as a cushion if there were unexpected variations in the operating conditions or device characteristics.

At this time the design of a high power, high quality audio amplifier presented an interesting technical challenge, in the absence of any high voltage, high power PNP transistors which could be used in conjunction with NPN power transistors in an output stage having complementary symmetry. The general philosophy used by Quad seemed to offer an answer to this problem, and I have shown in Figure 8.10b a layout for a Q/C triplet which I tested for use in a high power amplifier. Under DC or LF conditions, the two halves of this triple were virtually identical, and the optimum quiescent current ($\approx$100mA) was also the same for both emitter-follower groups. Used in the output stage of an experimental amplifier design, this output configuration gave a distortion figure at low output power levels which was less than my then ability to measure it – at the time my test bench THD meter had a lower measurement limit of about 0.05% over the range 100Hz–10kHz – and the amplifier did not appear to exceed this threshold value over the output power range from 10mW to 30 watts.

Another method for improving the symmetry of the output stage was suggested by Shaw (Shaw, I.M., *Wireless World*, June 1969, pp. 265–266), and shown in Figure 8.11a. Baxandall did an analysis of this layout (Baxandall, P.J., *Wireless World*, September 1969, pp. 416–417) and proposed a rather more straightforward way of achieving the same end, using the circuit layout shown in Figure 8.11b. Unfortunately he did not extend his analysis to show a fully worked out amplifier design based on his analysis. I was attracted by the simplicity of this approach, in which the diode, D1, simulated the effect on the driver transistor of the base-emitter junction of the lower output transistor. The effect of this diode in imitating the missing output transistor

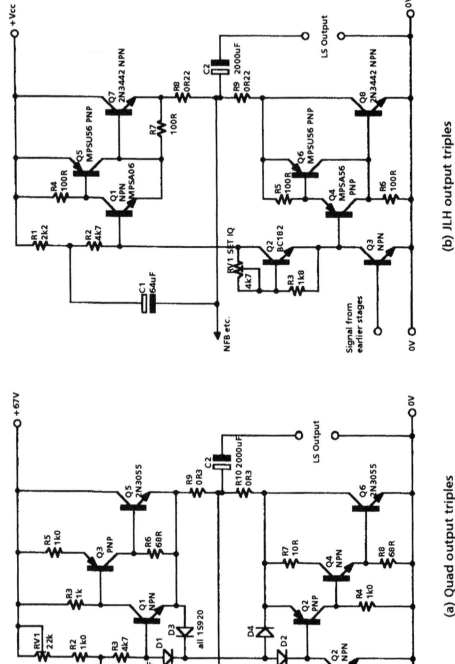

Figure 8.10
Improved Q/C output stages

(a) Quad output triples

(b) JLH output triples

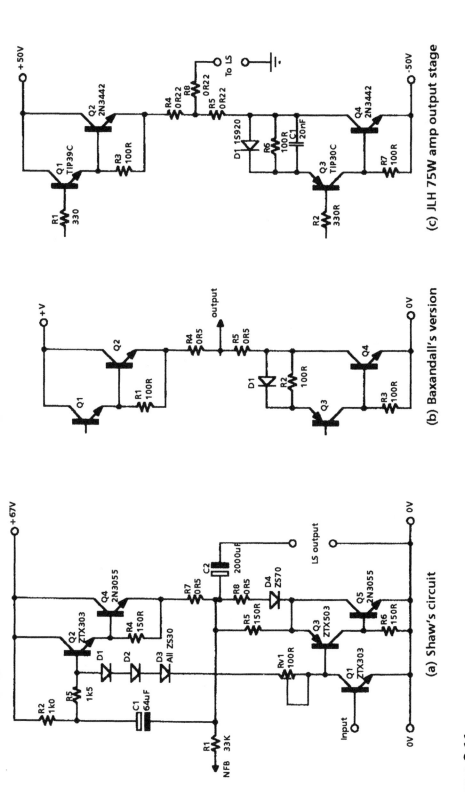

Figure 8.11
Improved Q/C output stages

(a) Shaw's circuit

(b) Baxandall's version

(c) JLH 75W amp output stage

junction could be improved, especially at higher frequencies, by adding a capacitor, C1, across this diode to simulate the output transistor forward junction capacitance, as shown in Figure 8.11c. I adopted this circuit for the output layout of a 75 watt amplifier design, eventually published in *Hi-Fi News* (Linsley Hood, J.L., *Hi-Fi News and Record Review*, November 1972, pp. 2120–2123) in what became a very popular constructional project. It is easy for an author to think favourably of his own designs, but my personal feeling, then and now, was that with this design, and others of similar quality which were then being offered, junction transistor audio amplifiers had come of age, and that their users need not feel that something had been lost for ever with the passing of thermionic valve operated designs. I have, for the record, shown the circuit of this 75 watt amplifier in Figure 8.12.

Although there are one or two innovations in this circuit, the design is fairly straightforward, and consists of an input long-tailed pair stage, with a junction FET used to provide a very high dynamic impedance constant current source tail to improve the emitter signal transfer between Q1 and Q4. PNP transistors are used in this stage so that Q5, the main voltage amplifying stage, could be a high voltage NPN device with good HF response. The stage gain was increased by the use of a DC bootstrap circuit (Q3,R3,R6) as the load for Q1. This also gives a low drive impedance for Q5, which also helps to maintain the stage gain. The voltage drop which is developed across an amplified diode, Q6, due to Q5 collector current, is used to provide the forward bias (about 3V) needed to make the output transistor groups operate at the best point on their combined push-pull transfer characteristics. Q6 is mounted on the output transistor heat sink to provide a measure of thermal compensation for the quiescent operating current, and helps to maintain this at the desired level (≈ 100mA).

A bootstrapped load resistor (C8,R13) is used to increase the dynamic impedance of R13, the collector load of Q5. The operation of the Shaw/Baxandall technique used to increase the symmetry of the output Q/C transistor layout has already been described and illustrated in Figure 8.11. HF stability for all likely combinations of reactive loudspeaker loads is ensured by the main, dominant lag capacitor, C9, connected between Q5 collector and Q4 base – in which position it does not contribute to slew rate limiting or slewing induced distortion, an immunity which is assisted by the input low pass network R2/C2. Since a large amount of NFB (approximately 46dB) is employed to maintain a very low level of distortion over the whole available output power range, the feedback loop characteristics are tailored by the HF step networks R9/C6, R3/C3, R4/C4 and the output Zobel network C14/R31 so that the loop phase characteristics are satisfactory. Typical performance figures for the design shown in Figure 8.12 are: output power 75 watts into an 8 ohm load, bandwidth 15Hz–20kHz, (upper end set by R2/C2), THD <0.01% at all power levels below the onset of clipping, unconditionally stable into all combinations of load impedance or reactance.

Class A Operation

The other option which was open to the circuit designer, even in the absence of satisfactory PNP power transistors, was to operate the amplifier in class A, a

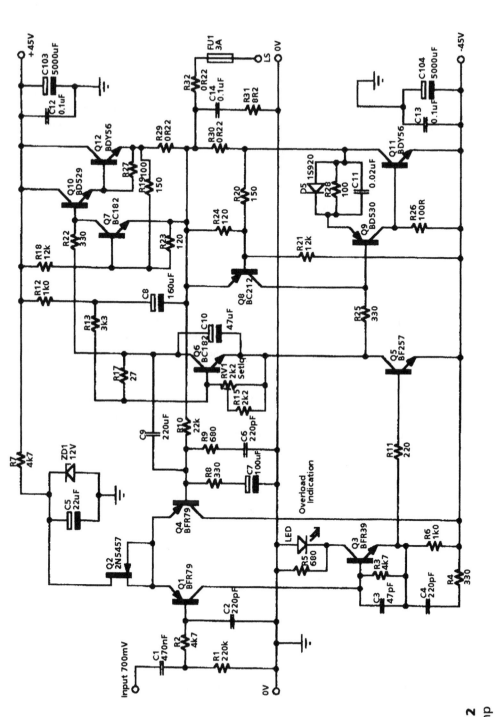

Figure 8.12
75 watt amp

possibility which, as shown in Figure 8.13, could be realised using only NPN polarity output devices. Since this is not a push-pull layout, crossover distortion cannot occur, but, since it is not a push-pull system, the output power available is limited, as in any other single ended layout, by the choice of the output stage operating current, and this, in turn, is limited by the permissible thermal dissipation of the output transistors. With reasonably efficient loudspeaker units, the bulk of normal listening would take place at output power levels which did not exceed a watt or two and the possible output power from such a class A system would be entirely adequate.

I had designed and built this amplifier for my own use, like all of my audio circuit designs up to that time, and I only offered it for publication because the use of output transistors in class A had become, at that time, a matter of topical interest, principally because a commercial amplifier using this principle, made by J.E. Sugden Ltd, had attracted very favourable reviews in the Hi-Fi press, who applauded its freedom from transistor sound.

The structure of the circuit shown in Figure 8.13 is very simple, with Q1 acting as a grounded emitter amplifier stage, with Q2 as an active collector load, driven in phase opposition to Q1 by Q3. The loop gain of the amplifier is increased by bootstrapping the load resistor for Q3 by C1. Because the transition frequency of the output transistors is of the order of 4MHz, whereas those of Q3 and Q4 are in the 400MHz range, the circuit has an in-built dominant lag in its loop NFB characteristics. This ensures that the loop gain has fallen below unity before the loop phase angle reaches 180°. No additional HF compensation networks are therefore necessary to ensure complete loop stability, even with reactive loads.

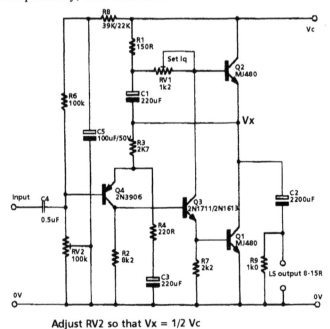

Adjust RV2 so that Vx = 1/2 Vc

Figure 8.13
JLH 10 watt class 'A' WW 4/69

Fully Complementary Designs

With the continuing development of epitaxial base and similar structures for silicon transistors, PNP power transistors became more readily obtainable, although initially in relatively limited voltage ratings, at prices which approached those of existing NPN power devices. This provided an incentive to the circuit designers to provide amplifier systems which took advantage of this new technology, and offered the possibility of reducing low signal level crossover distortion to a level where it would no longer be audibly detectable. Two of the circuit designs which made use of this new-found freedom were due to Locanthi (Locanthi, B. N., *J. Audio Eng. Soc.*, July 1967, pp. 290–294) and Bailey (Bailey, A.R., *Wireless World*, May 1968, pp. 94–98). Of these, the Bailey circuit offered a somewhat lower level of THD and I have shown the circuit used in Figure 8.14.

Although NPN and PNP power transistors were nominally exact equivalents of one another, in reality there were significant differences between these structures which lessened the symmetry of the amplifier circuits built around them. Of these differences, the most obvious was that the current carrying majority carriers were electrons, in the case of the NPN devices, and holes in the case of the PNP ones, and since electrons have greater mobility, performance differences show up at higher frequencies. The second difference, due to the nature of the emitter/base diffusion interface is that, although the makers quote identical safe operating area (SOAR) curves, PNP power transistors are nevertheless more prone to failure in use than NPN ones.

There are a number of interesting design features in the circuit of Figure 8.14, of which the most important, in terms of its influence on subsequent designs, was the output overload protection circuitry arranged around transistors Q5 and Q8. These are arranged to monitor both the output current from Q7 and Q10 (by means of the voltage drop across R25 and R26) and also the voltage present across Q7 and Q10. If the combination of voltage and output current is such to approach the secondary breakdown region of the maximum working limits of the device (see, for example, Figure 7.12), Q5 and Q8 will conduct, and limit the drive voltage applied to the bases of Q6 and Q9.

Bailey had also configured the circuit so that it operated between a symmetrical pair of voltage rails – anticipating the circuit configuration used in the so-called direct coupled layouts which subsequently became very popular (see, for example, Figure 8.12). However, he had chosen to retain a simple grounded base input amplifying stage, of which the inevitable V_{b-e} DC offset of Q1 was lessened by a bias current derived from the amplified diode circuit built around Q2, rather than the more straightforward input long-tailed pair layout of Figure 7.8a. Since the circuit of Figure 8.14 gave a DC output voltage offset under no-signal conditions, a reversible (AC working) electrolytic capacitor was required to isolate the loudspeaker from the amplifier output. Such capacitors are not commonly used, and are therefore expensive.

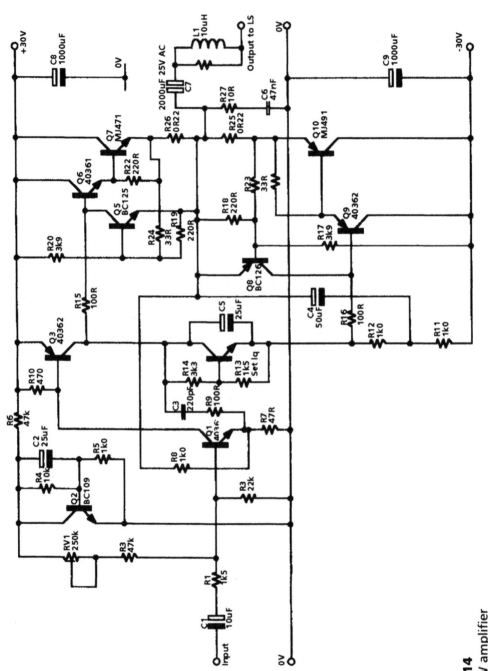

Figure 8.14
Bailey 30W amplifier

In the article in *Wireless World* in which he described this design, Bailey demonstrated, by oscilloscope traces of the relative V_{in}/I_{out} transfer characteristics, the fundamental lack of symmetry of the existing and widely used simple quasi-complementary layout, and showed the superiority of the fully complementary design in the way the harmonic distortion of the amplifier progressively decreased towards zero as the output power was reduced – behaviour which was typical of valve amplifiers operating in class A, but not found in most of the early class B (low or zero operating quiescent current) transistor designs. Bailey also showed the performance of his design when driven with a square-wave input signal and coupled to resistive or reactive loads. A lack of overshoot in, or significant distortion of, a square wave or similar type of signal, when the amplifier is caused to drive a reactive load, makes, I believe, an important contribution to good overall sound quality in an audio amplifier.

Gain Stage Designs

The gain stages between the signal input point and the output devices are normally operated in class A and are arranged to give as wide a bandwidth, as high a gain and as low a phase shift between input and output as is possible. To reduce the difficulty in keeping the final amplifier stable, when overall NFB is applied, the gain block is normally restricted to two amplifying stages, and to get as high a gain from these stages as is practicable, the collector load for the second stage will be arranged to have a high dynamic impedance. In the amplifier designs shown so far (Figures 8.4, 8.8, 8.12–8.14), this increase in the AC impedance for a given DC resistance has been achieved by bootstrapping the load resistor (by coupling its supply-line end by a capacitor connected to the output of the amplifier). In addition to increasing the AC impedance of the load resistor this also has the practical effect of increasing the possible output voltage swing which that stage can deliver. However, this technique is essentially that of applying positive feedback around the output stage. When this is an emitter-follower stage, or some similar arrangement, the gain will be less than unity and the amplifier will not be unstable. On the other hand, positive feedback has the effect of increasing both the stage gain and the distortion of the stage across which it is applied. In the present context, if that stage suffers from crossover distortion, the ill effects of this will be magnified by a drop in the dynamic impedance of the load resistor and a reduction in the driver stage gain at the crossover point. Modern design practice therefore tends to favour a high dynamic impedance load, such as a constant current source or the output from a current mirror, as the means of optimising driver stage gain. Typical arrangements of this kind are shown in Figures 8.15a–8.15c.

The layout of Figure 8.15a is a fairly typical long-tailed pair input stage, in which a constant current source, of one of the types shown in Figures 7.17 or 7.24, has been used as the tail in order to assure the integrity of the signal transfer between the emitters of Q1 and Q2. This approach is favoured in IC manufacture where it is easier, and less expensive of chip area, to manufacture an active device than a resistor – especially one of high value. Similarly, the gain of Q3 could be increased by replacing R2 with a further constant current source.

In the modification shown in Figure 8.15b, a current mirror, such as one or other of the types shown in Figure 7.25, has been used to combine the outputs of the two

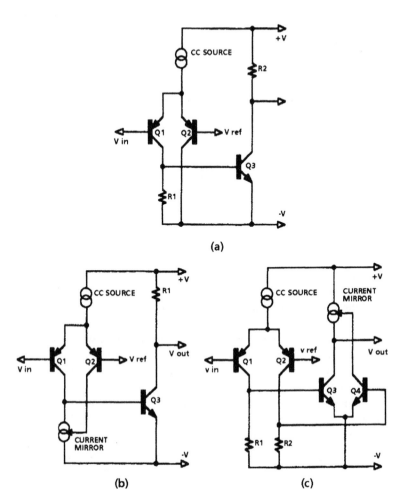

Figure 8.15
Typical gain stages

transistors of the long-tailed pair, which will substantially increase the input stage gain. Once again a constant current source could be used in place of R1, with a further increase in stage gain. As modified in this way layouts of the kind shown in Figure 8.15b form the basic structure of the bulk of both operational amplifier circuits (because it does not need any resistors) and of a large proportion of Hi-Fi audio amplifier circuitry.

In all of these layouts the polarity of the devices could be reversed (i.e. by substituting NPN for PNP devices, and vice versa) and other types of transistor, such as JFETs or MOSFETs, could be used, at the choice of the designer. Similarly, the gain can be further increased, especially at the higher end of the frequency band, by connecting a cascode transistor, of one of the forms such as are shown in the layouts of Figures 7.7 or 7.15, between its collector and the collector load.

An interesting further development of this idea is shown in Figure 8.15c, in which the current mirror, used to combine the outputs of the two antiphase signal streams, is transferred to form the collector load of the second gain stage transistor, Q3. This idea appears to have been originated by National Semiconductors, and has been used in several of their IC designs, such as the LH0061. It has also been adopted by Hitachi as the basis of an audio amplifier design (Hitachi Ltd, *Power Mosfet Application Manual* (1981), pp. 110–115), having the desirable qualities of symmetry and a very high gain from just two stages.

HF Compensation Techniques

This is the somewhat misleading term which is given to the adjustment of the amplifier gain and phase characteristics, as a function of frequency, so that when overall loop feedback is applied the amplifier remains stable – ideally with a wide margin in terms of the gain or the phase angle which exists between the working condition of the amplifier and the onset of instability. While a two gain-stage amplifier of the kind shown in Figure 8.15 would most probably be stable if an NFB signal was returned from the output to Q2 base, by way of some suitable network, if a push-pull output emitter-follower pair, of the kind shown in Figures 8.12 or 8.14, were to be interposed in the feedback path, the loop phase shift would approach 180° at some upper or lower frequency at which the loop gain was equal to, or exceeded, unity, and the amplifier would oscillate.

With direct-coupled circuits the LF phase shift will not exceed a safe level, so the problems of loop instability are confined to the HF end of the signal pass-band, and it was (and is) customary to achieve the necessary HF loop stability by imposing a single-pole, dominant lag characteristic on the gain/phase relationships of the system by connecting a small capacitor (C_{fb}) between the collector and base of the second amplifying transistor (Q3) in Figure 8.16a, since this arrangement gives the best performance, in terms of THD, at the high frequency end of the pass-band. However, this approach leads to the problem that it imposes a finite speed of response to any rapidly changing input signal while C_{fb} charges or discharges through its associated base or collector circuits – mainly being limited by the collector current of Q1. This effect is illustrated in Figure 7.27.

If a composite input signal which includes some rapid rate of change of input voltage level is applied to the input of the amplifier it is possible that the input device (Q1) will be driven into cut-off or saturation because no compensating feedback signal has yet arrived at the base of Q2. If this happens, there will be a complete loss of signal during this period because Q3 will be paralysed while the charge on C_{fb} is returning towards its normal level. This problem was described by Otala in a published paper (Otala, M.J., *J. Audio Eng. Soc.*, 1972, No. 6, pp. 396–399), and he coined the term Transient Intermodulation Distortion to describe the audible effects of this type of malfunction. A simpler description, suggested by Jung (Jung, W.C., *Hi-Fi News and Record Review*, November 1977, pp. 115–123), is slewing-induced distortion (or slew rate limiting), and this effect can be seen clearly on an oscilloscope connected to the output of an amplifier when a suitable input signal is applied.

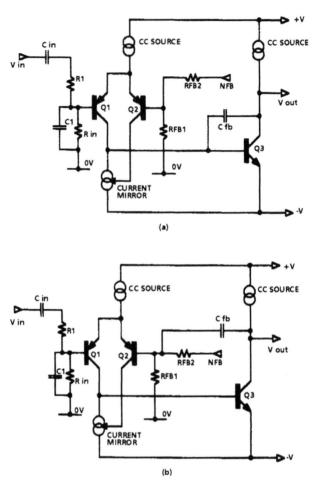

Figure 8.16
HF compensation methods

This type of problem is not an inevitable consequence of dominant-lag type HF compensation since there are ways of avoiding it (Linsley Hood, J.L., *Hi-Fi News and Record Review*, January 1978, pp. 81–83). Of these, the simplest is just to introduce an RC low pass network at the input of the amplifier to limit the possible rate of change of the input voltage – as R1/C1 in Figure 8.16. A better approach is to include the whole of the amplifier gain stages within the bandwidth limiting system, as used, for example, by Bailey (C3 in Figure 8.14), and illustrated in Figure 8.16b. Then, provided that the possible rate of change of the collector voltage of Q3 (which is determined by its collector current and the circuit capacitances associated with its collector) is faster than the rate of change permitted by R1/C1 (which is within the control of the circuit designer) slew rate limiting will not occur.

Although the conventional scheme shown in Figure 8.16a is better from the point of view of THD in the 10kHz–20kHz part of the pass-band, it is much inferior in respect of the normal square wave into reactive load type of test, and it has always seemed unwise to me to choose a design approach in which an inaudible improvement – from 0.1% to 0.02% THD at 20kHz – has been bought at the cost of worsening the (almost certainly audible) transient error from, say, 2% to 40%, as measured, for example, at 10kHz in the manner shown in Figure 8.17. In this, the amplifier, operating with a simulated reactive load, is fed with a good quality square wave, and an amplitude and time-delay corrected square waveform, derived from this, is subtracted from the amplifier output. Ideally the residue should be zero, and the closer the amplifier output approximates to its input waveform under reactive load conditions, the better it will probably sound.

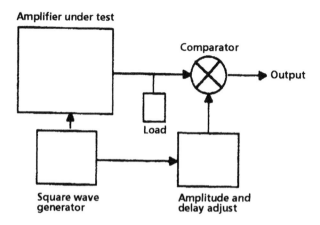

Figure 8.17

Symmetry in Circuit Layout and Slewing Rates

A residual problem with any system in which there is stray, or other circuit capacitances, is that the maximum possible slewing rates will not necessarily be the same for a negative-going or a positive-going signal excursion. This is because the outputs of circuits do not necessarily have the same ability to source or sink current, so there must inevitably be differences in the rate in which any associated capacitances can be charged or discharged. To take the case of the very simple amplifier layout shown in Figure 8.15a, if there was some capacitance between its output and the 0V line this capacitance could be discharged very rapidly if Q3 were turned fully on, but would, perhaps, only charge up again, towards the positive line, at a slower rate, which would depend on the value of R2. This problem would be worsened if a constant current source were used instead of a resistor, as in the, apparently much preferable, circuit of Figure 8.16a. This prompted some designers, such as Bonjiorno (*Audio*, February 1974, pp. 47–51) and Borbely (*Audio Amateur*, February 1984, pp. 13–24), to propose circuits of the form shown in Figure 8.18, in which two amplifier blocks of the kind shown in Figure 8.16a are coupled together as a mirror-image pair. The only drawback with this type of layout is that there is normally rather greater difficulty in achieving a stable quiescent current in the output transistors – a thing which is very desirable in any class AB output stage for optimally low levels of crossover distortion.

Stability of Output Current

By the early 1970s audio amplifiers operating in class AB – by which I mean those in which it is intended that a small residual current will pass through the output devices under zero signal conditions – had mainly achieved very high standards of performance, although the one remaining disadvantage which they all shared was that the preferred setting of the quiescent current could be quite critical, and would need to be set up on the test bench for each amplifier as the final step of its assembly. Moreover, there was no guarantee that this I_q setting, when correctly adjusted (ideally while monitoring the output with an oscilloscope and a distortion meter), would remain at the chosen value, or even that this chosen value would still be the correct one, during the aging of the circuit components, or as the ambient temperature of the amplifier changed.

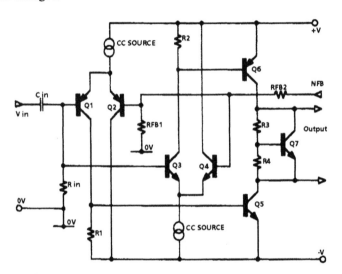

Figure 8.18
Fully symmetrical gain stage

A number of design proposals were offered as a means for ensuring the stability of the quiescent current, but, in general, these all suffered from disabling flaws in their design, so that, in practical terms, the designers were left with the options of trying to ensure quiescent current stability in the face of operating changes or to choose a design approach in which the actual quiescent current value was not particularly critical. An example of the latter approach was my 15–20 watt class AB design of 1970 (Linsley Hood, J.L., *Wireless World*, July 1970, pp. 321–324).

Various circuit arrangements had been adopted to minimise the effect of changes in the temperature of the output devices, but one bold approach, due to Blomley (Blomley, P., *Wireless World*, February 1971, pp. 57–61 and March 1971, pp. 127–131), is shown, in a slightly simplified form, in Figure 8.19. In this the output devices, a complementary pair of silicon planar transistors, are permanently biased into conduction, and the input signal, after amplification, is chopped into two halves by a pair of switching transistors (Q1 and Q2), and these halves are then passed to the output transistor triples for reassembly into an enlarged and power-augmented version of the original signal. The snag, of course, is that a correct forward bias must now be chosen for the small-signal switching transistors, which merely moves the problem of choosing and maintaining the operating current away from the output transistors and back to the earlier switching stage.

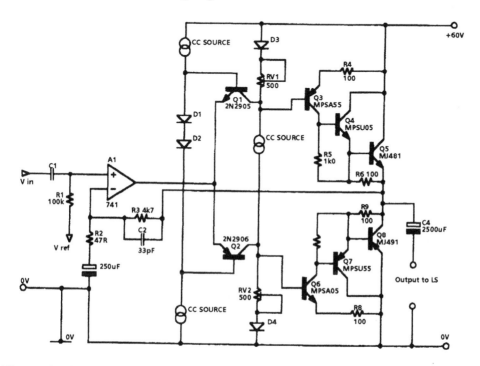

Figure 8.19
The Blomley amplifier

A typical example of bipolar transistor operated audio amplifier design of the mid-1970s, incorporating many of the contemporary design features, is shown in Figure 8.20. This has an excellent performance both in terms of its THD (better than 0.01% at 50 watts output over the frequency range 30Hz–15kHz) and square wave into reactive load (no significant overshoot when coupled to an 8 ohm load in parallel with capacitor values from 1nF to 4μF). Q2 and Q3 form a constant current source for the input long-tailed pair (Q1/Q4) with a preset potentiometer (RV1) connected between its emitters to allow the DC offset at the IS output terminals to be reduced to a very low level. Q5 and Q6 form a current mirror load for the input stage and Q10 acts to protect Q7 from an excessive input drive signal.

Q8 and Q9 form a constant current load for the main gain stage transistor (Q7) which drives the output devices on either side of the forward bias generating network (RV2/C7) through hang-up prevention resistors (R11/R12). The output transistors are connected as compound emitter-followers because this arrangement gives an output which more nearly equals the input (because of its higher loop gain) and because it offers a lower output impedance. This arrangement also has the advantage that the base-emitter junctions (which will become hot in use) are not directly involved in determining the best forward bias setting – this depends on Q11 and Q13 which have a much lower output current level. Q9 senses the ambient temperature, and adjusts Q7 collector current and the voltage drop across RV2 as required.

Output overload protection could be by means of an output fuse, as shown, or an output cut-out relay, or a current limited power supply. The use of Bailey–type output protection circuitry had fallen from favour at that period because it was thought to operate prematurely during high level signals, especially at higher audio frequencies where the impedance of many commercial LS systems may fall to a low level, and be seen as an apparent output short-circuit.

Although there were still areas in which improvements could be made, a comparison between the performance given by the late 1950s design of Figure 8.3 and that of the mid-1970s design shown in Figure 8.20 indicates the extent of the progress made.

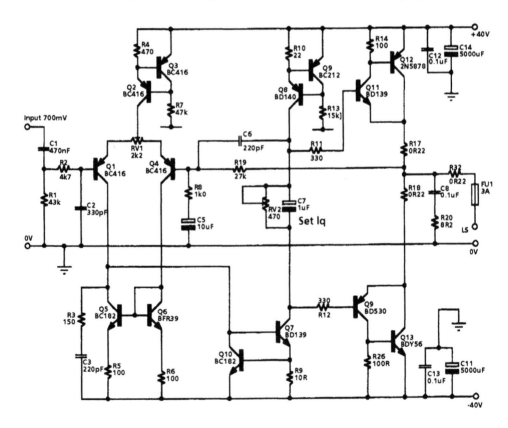

Figure 8.20
1970s 60 watt amplifier

CHAPTER 9

CONTEMPORARY POWER AMPLIFIER DESIGNS

In the last chapter I looked at the evolution of audio amplifiers, based on bipolar junction transistors, from early transformer coupled designs, having relatively unsatisfactory electrical characteristics, to comparatively sophisticated circuit layouts having a performance, both on paper and in the ears of any unprejudiced listener, which was fully comparable to that of the best valve operated designs of former years. This, I know, will be thought to be a contentious claim by many of those who base their own technical assessments either on what they are told by those whose experience and acoustic judgements they value – perhaps because they are well known in Hi-Fi circles – or on the opinions they themselves have formed during listening trials, a complex topic I propose to consider later in this chapter.

Remaining Design Problems

At the state of the art in the early 1970s, the design problems which remained – and these were only small imperfections or minor difficulties which did not usually inconvenience the user – were:

- the need, initially, and perhaps also from time to time, to adjust the quiescent current of the output push-pull stage to obtain the best practicable performance;

- the need to ensure that this quiescent current value remained at its optimum setting throughout the aging of circuit components or changes in junction temperatures of the transistors;

- the need, especially in the direct coupled designs, where there was no DC blocking capacitor between the amplifier output and the loudspeaker unit, to prevent possibly damaging DC voltages appearing across the LS terminals in the event of an improper input signal or a component failure;

- the need (since all sensibly designed amplifiers will employ negative feedback) to make sure that the stability margins of the feedback loop were sufficient to allow electrically awkward LS loads to be driven without significant waveform distortion or amplifier instability;

- the need to evolve circuitry and components which would allow progressively higher output powers to be provided for use with increasingly inefficient LS systems. In practice, changes in outlook and expectations have meant that while a design having a 10 watt output would have been considered entirely adequate in the late 1950s, and one of 30 watt capacity quite sufficient in the 1960s, one of 150–250W output would not be thought to be needlessly over-powered in the 1990s;

- the need, because of a change in the majority usage of Hi-Fi equipment from classical music to various kinds of high sound-level popular styles, for all the electronically operated audio units to handle high level signal inputs without overload, or when such signal overload did occur, that the circuitry involved should handle this overload gracefully.

Power MOSFETs vs. Bipolar Junction Power Transistors (BJTs)

The major change in component availability since the late 1970s has been the growing use of power MOSFETs. The construction and general characteristics of these devices in T, D, V and U versions were explored in Chapter 7 (cf. Figure 7.22), and some of their advantages and disadvantages were considered. Where they are used in audio circuitry, such devices are normally employed as the amplifier output transistors, almost always as a complementary, N-channel/P-channel, push-pull pair, since the electrical characteristics of these two MOSFET types are much less dissimilar than those of an NPN/PNP pair of junction transistors. However, in addition to the high- and medium-power devices, small-signal versions, mainly of T-MOS devices, are also manufactured, at voltage ratings up to 600V and current ratings up to 0.7 amperes, and permitted dissipation ratings from 300mW to 1W, in the case of the TO–237 encapsulated N-channel transistors. Similar components, though only at present at working voltages up to 450V (V_{d-s}), are obtainable in P-channel versions.

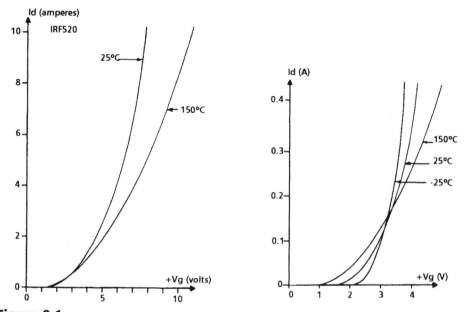

Figure 9.1
MOSFET characteristics

MOSFETs have a number of advantages and drawbacks in comparison with power BJTs. Of these the first snag is that power MOSFETs are up to three times more expensive to buy than comparable power BJTs, and the relative cost of small-signal MOSFETs may be even higher. This extra expense is mainly a function of the chip size, which determines the number of devices which can be fabricated, at the same time, on a given size wafer of single crystal silicon, and, following from this, the effects of the relative economies of scale in mass produced components. MOS devices having a low conducting resistance (R_{DSon}) because a large number of channels have been connected in parallel, will obviously require a larger chip size, as will devices having higher working voltages – which require wider spacings between the diffusion zones. It is also claimed that MOSFETs are less linear than power BJTs when used as output source-follower pairs. However, this is only marginally true, as can be seen from the data given in Table 9.1.

Frequency Response of MOSFET Devices

The major feature of the MOSFET, which can be both an advantage and a drawback, depending on the application, is that it is very much faster in responding to an input voltage change than any BJT having comparable voltage and current ratings. One consequence of this good HF response is that, in an incompetently designed MOSFET circuit, the devices may burst into oscillation, at an extremely high frequency, as soon as the supply rail voltage is applied. Since such inadvertent oscillation may occur at a frequency which is well above the HF bandwidth of the oscilloscope which is being used to monitor the circuit, the hapless designer may only become aware that he has a problem when he finds that his output devices have become very hot and have consequently expired, for reasons which may not be apparent. The simplest answer to this particular problem is to connect a gate stopper resistor, of appropriate value (say, 150–1000Ω), in the MOSFET gate lead, as close as possible to the gate pin, and to ensure that the layout of the gate, source and collector leads is not such as to encourage inadvertent Colpitts, or Lecher-line–type, oscillation.

In a well-designed circuit, the MOSFET is likely to prove more robust in use than the BJT since it does not suffer from thermal runaway, and its operating characteristics are free from any secondary breakdown region, see Figure 7.20 vs. Figure 7.12. This saves space and expense by avoiding the need for elaborate output transistor protection circuitry.

A complementary pair of these is also, as noted above, much more truly symmetrical – particularly at high frequencies – than an NPN/PNP pair of junction transistors. A further factor in favour of the MOSFET is that the operating current is carried by majority carriers (electrons in an N-channel device and holes in a P-channel one), and it is therefore free from the hole-storage defects which can contribute to the sluggish turn-off characteristics of any BJT, following a condition in which it has been caused to turn hard on, as might happen during a brief signal overload.

However, the major quality which endears MOSFETs to the audio amplifier designer is its high effective f_T, which can be in the 100MHz region. The speed of response of any MOS device is mainly controlled by the time it takes the capacitance inherent in the gate electrode to charge up to its target voltage through the finite resistance (or current limit) imposed by the driver circuitry. (Since the channel current is controlled by the gate voltage – in an enhancement mode MOSFET, which means all the normal power devices – until this voltage appears on the gate, no channel current will flow.) This means that the designer has to hand, in the form of the gate stopper resistor, a simple tool for controlling the effective f_T of the transistor. This is also true of the power BJT, but in this case the normal problem is that the f_T is not as high as one would wish, rather than that it is too high.

Finally, since monocrystalline silicon has a negative temperature coefficient of conductivity, MOSFETs can show a reduction in drain current for an increase in temperature – at least at larger values of drain current – and this removes the likelihood of thermal runaway under heavy load. It also implies, as seen in Figure 9.1, that there will be some value of bias voltage/output current where the quiescent current setting will be nearly independent of temperature; a fact which is convenient in push-pull output stage design. I have shown, for comparison, the influence of junction temperature on the drain and collector currents of MOSFETs and BJTs in Figures 9.1 and 9.2, and I have also, for clarity, expanded the low current region of Figure 9.1 to show on a larger scale the effect of the negative temperature coefficient of the MOSFET drain current.

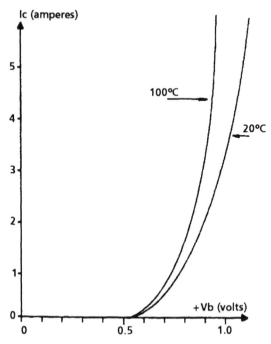

Figure 9.2
Power BJT characteristics

MOSFET Linearity

One of the criticisms levelled at MOSFETs, often made I suspect by those engineers who have found experimental difficulties in obtaining stable operation, is that they are much less linear than junction power transistors, when used as output push-pull pairs of emitter- or source-followers. These claims have generally been based on the results obtained from computer circuit simulations of the push-pull transfer characteristics (cf. Self, D., *Electronics World + Wireless World*, May 1995, pp. 387–388), although comparisons between actual audio amplifier circuits using MOSFETs or BJTs as the output devices (cf. Toshiba, *File Note X3504*, March 1991) have shown the converse: that amplifier circuits using MOSFET output devices have a lower distortion than those using BJTs.

It is inevitable that any push-pull pair of devices whose input voltage/output current relationships fall short of an ideal straight-line graph will introduce some distortion into the output signal, whatever bias point is chosen, and there are big differences between the BJT and the MOSFET in this respect, as shown in Figures 7.3 and 7.18. These differences will influence the relative distortion introduced even by an optimally biased output push-pull pair, and I have shown the comparative results from tests using four such common circuits in Table 9.1.

Table 9.1 Distortion of output push-pull pair layouts

Configuration	I_q	THD (5kΩ load)	THD (10R load)
(A) Darlington pair	60mA	<0.02%	2.2%
(B) Compound E/F	60mA	<0.02%	1.15%
(C) MOSFET s-follower	100mA	0.02%	2.1%
(D) BJT/MOSFET pair	100mA	0.015%	1.0%

Note. These measurements were made at 20kHz, at a 5V rms output voltage and a 10Ω load. The circuits used are shown in Figure 9.3, and the 20kHz measurement frequency was chosen to show up any differences between nominally complementary devices due to dissimilarities in their HF response.

The comparative measurements shown above are of the output push-pull pairs on their own, unencumbered by ancillary circuitry, and freed from any complicating effects due to the HF compensation techniques which are necessary in any feedback (voltage) amplifier circuit. The layout of Figure 9.3a is that of a conventional Darlington pair output arrangement – very commonly used in solid state audio systems. That of Figure 9.3b is the rather better compound emitter follower system, in which R2 and R5 have been included to make the setting of the required bias voltage rather less abrupt. Figure 9.3c employs a simple MOSFET complementary-pair source follower output stage –

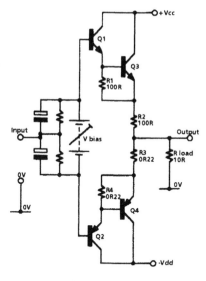

(a) Darlington pair

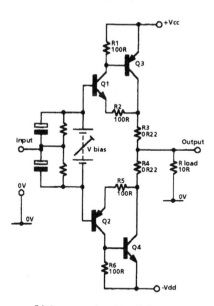

(b) Compound emitter–follower

Figure 9.3a/b
Output push-pull pairs

which is in essence the same as that of the Darlington pair circuit of Figure 9.3a except that because the MOSFET has a relatively high input impedance, no impedance transforming driver transistors, such as Q1/Q2 in Figure 9.3a, are needed.

(For the avoidance of possible spurious HF oscillation in the circuit of Figure 9.3c not only must gate-stopper resistors (R1/R4) be used, but the MOSFET source electrodes

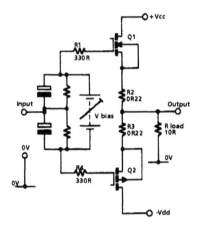

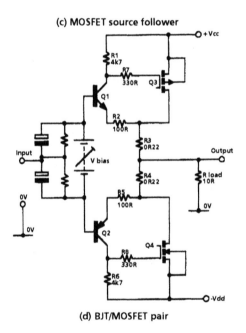

(c) MOSFET source follower

(d) BJT/MOSFET pair

Figure 9.3c/d
Output push-pull pairs

must be protected from capacitative loads. The small inductance ($\approx 1\mu H$) of the 0R22 wirewound source resistors serves this purpose adequately.)

The circuit of Figure 9.3d is that of the compound emitter follower in which the output junction transistors have been replaced by complementary power MOSFETs. In this case gate over-voltage protection may be provided simply by connecting 10V Zener diodes across R1 and R6. I used this output stage arrangement as a recommended improvement to a previously published audio amplifier design of my own, in which I had originally used monolithic Darlington BJTs (Linsley Hood, J.L., *Hi-Fi News and Record Review*, December 1980, pp. 83–85).

In view of the claims for high distortion levels in MOSFET output stages, the surprising thing about the results of the trials detailed in Table 9.1 is how similar in practice the distortion levels of the BJT and the MOSFET circuits were (a factor in determining overall THD levels, as seen below). In both cases the compound emitter–follower arrangements were superior in performance to the straight Darlington pair, or the MOSFET source–follower circuit, by an approximate factor of two. A second surprising feature, observed and used by Sandman (Sandman, A.M., *Wireless World*, September 1982, pp. 38–39), is that with a high impedance output load (in the tests made, this would have been about 5kΩ) the residual distortion introduced by a simple emitter- or source-follower is very low indeed and, I observed, relatively unaffected by the actual I_q value chosen, over quite a wide range of settings. To be able to obtain such a low level of crossover distortion and such an uncritical setting for the I_q value would remove one of the major problems in solid state audio power amplifiers, and this philosophy prompted Sandman to propose an audio power amplifier design in which an ancillary output circuit was used to provide a very high impedance load for the basic amplifier stage.

g_m Values for BJTs and MOSFETs

In a bipolar junction transistor there is a predictable relationship between g_m and collector current, as shown in the equation. This is only affected to a minor extent by the device geometry:

$$g_m = I_c(q/kT)$$

where q is the charge on the electron ($1.60 \cdot 10^{-19}$), k is Boltzmann's constant ($1.38 \cdot 10^{-23}$), and T is the absolute temperature of the junction – giving a room temperature value for g_m of 38.9S/A. In the case of both the junction FET and the MOSFET the drain current vs. gate voltage characteristics have theoretically a square–law characteristic, defined, in the case of the MOSFET, by the relationship:

$$I_d = (V_{GS} - V_T) \approx (\mu_e C_0 Z)/2L$$

where V_{GS} is the gate-source voltage, V_T is the threshold (turn-on) voltage, μ_e is the electron mobility in the channel region, C_0 is the capacitance, per unit area, of the gate oxide layer, L is the channel length and Z is the channel width. This relationship can be simplified to the approximation:

$$g_m \approx 1/(R_{DSon})$$

which would give a g_m in the range 5–20S for a typical 100V N-channel TMOS power transistor. However, in both the JFET and the MOSFET the electrical characteristics are greatly influenced by the internal structure of the transistor – an area in which development work is continuing (c.f. the recently introduced Philips Trench-MOS devices with their very low values of R_{DSon} and their very high and linear g_m) – and most manufacturers use their control of this structure to manufacture components which are somewhat more linear than the square-law behaviour predicted by the formulae.

Quiescent Current Stability in Push-pull Output Systems

The problem of setting and maintaining, once set, the optimum forward bias for a class AB or class B output stage* has been a problem with low dissipation solid state audio amplifiers ever since such things were made. There are, basically, three solutions to this problem:

- To choose a circuit structure (such as, for example, a MOSFET complementary source-follower output pair) in which the actual value of the quiescent current is not particularly critical, so that some mid-point value can be chosen in the expectation that the resulting performance will not be too far away from the optimum; in practice, this will probably imply a fairly large value for I_q. (Typical power MOSFETs show a region on their I_D/V_G curves, usually for drain currents in the region 100–200mA, at which there is little change in I_D over the temperature range –50°C to +150°C.)

- To use some circuit technique, such as those illustrated in Figure 9.4, that will provide adequate compensation for changes in the junction temperatures of the output devices (essential in the case of bipolar power transistors) – for which proposed methods range from the simple Bailey amplified diode maintained in thermal contact with the output heatsink to the elaborate custom-built IC used by Pioneer to regulate the output device quiescent current shown in Figure 9.5.

The circuit shown in Figure 9.4a is a simple elaboration of Bailey's amplified diode which I used in my 1972 75W amplifier (Linsley Hood, J.L., *Hi-Fi News and Record Review*, November 1972, pp. 2120–2123), while that of Figure 9.4b makes use of the Williams–type constant current source built around Q1/Q2 to control the current flowing through R3, and hence the forward bias on the output devices. The layout of Figure 9.4c is a further elaboration of the Bailey amplified diode arrangement, due to Self (Self, D., *Electronics World + Wireless World*, October 1996, p. 755), in which a further diode is used to increase its thermal sensitivity.

These terms refer to the output stage biasing conditions, in that class AB operation implies that there is some residual current flow under no-signal conditions, whereas the class B designation implies that there is no such quiescent current, so that, on signal, either one or other of the two output halves will be cutoff, alternatively, on each half cycle of the signal, as the output signal voltage swings up or down. In practice, since this type of operation would lead to audible discontinuities around the near-zero signal regions, the vast majority of transistor operated audio power amplifiers, whose designers have sought the best compromise between quality of sound and evolution of heat, will operate in class AB

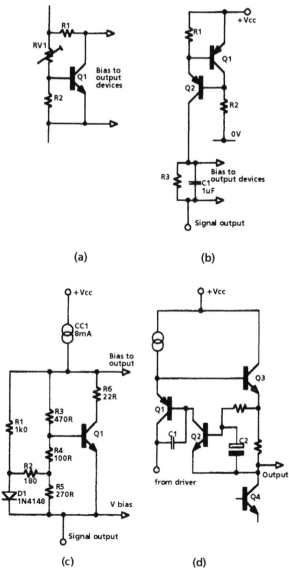

Figure 9.4
Output stage bias systems

Other I_q Control Techniques

A number of circuits have been proposed in which the emitter/collector currents of the output transistors are monitored, and used to regulate the forward bias voltage so that the desired quiescent current is held to some constant value. In general, this type of control arrangement is only satisfactory with a system biased into class A operation, such as that described by Nelson-Jones (Nelson-Jones, L., *Wireless World*, March 1970, pp. 98–103), and shown in Figure 9.4d. Even then, a large value capacitor (as C2 in the diagram) is needed to remove the low frequency signal components from the I_q control voltage.

In all cases where BJTs are used as the output devices it is essential that the sensing transistor (Q1 in the diagrams of Figure 9.4) should be in as close thermal contact with the output transistors as possible. Self even proposes gluing the sensing transistor to the case of one or other of the output transistors (Self, D., *Electronics World + Wireless World*, May 1996, p. 412) in an attempt to reduce the time lag between the output transistor heating up during a high signal-level programme passage – which will cause an increase in the output quiescent current of the BJTs – and this temperature increase being sensed, or the corresponding time lag between the output BJTs cooling down again, during a quiet passage in the programme – which will reduce their quiescent current setting – and the sensing transistor registering this change in heat evolution. Of these two periods of discontinuity between the output transistors heating or cooling and that fact being registered by the sensing transistor, it is the second which is more important to the performance of the amplifier since the nature of the crossover distortion – particularly in a BJT output stage – becomes less acceptable at quiescent current settings below the optimum value.

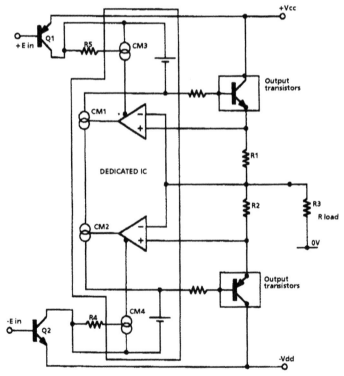

Figure 9.5
Pioneer Iq control circuit

In the Pioneer M–90(BK) 250 watt amplifier, which has groups of eight parallel-push-pull connected BJTs in the output stage of each channel, a purpose-built IC, having the circuit layout shown schematically in Figure 9.5, is used to sense the temperature of the amplifier, the quiescent current of the output devices and the signal level, and to anticipate and control variations in the quiescent current setting of the output devices. A further system for quiescent current control in an output pair of current driven MOSFETs was proposed by Van de Gevel (Van de Gevel, M., *Electronics World + Wireless World*, February 1996, pp. 140–143), and is shown, in greatly simplified form, in Figure 9.6. In this the source/drain current of the output MOSFETs (Q3/Q4) is converted into voltages across R1 and R2, and these voltages are compared with fixed reference voltages by Q5 and Q6. The sum of these voltages – converted into a current by Q5 and Q6 – is then used, via the two current mirrors, to modulate the collector currents of Q1 and Q2, and hence the bias voltages applied to the output MOSFETs. The purpose of this arrangement is that the effect of input signals which are applied differentially to the output devices is partially cancelled by subtraction, and is not used to diminish the I_q level, whereas those changes which occur in both devices (i.e. due to changes in static I_q) are added in the control system. It is claimed that this also has the effect of ensuring that neither output transistor can be cut off at any point of the input signal.

However, although the preceding circuit layouts partially solved the problems of variable quiescent current and the need to set this, at least once, during manufacture, or use, it still did not provide a design basis for an amplifier which required no setting up. The first really satisfactory circuit arrangement which remedied this problem was introduced in 1975 by the Acoustical Manufacturing Co. (Quad).

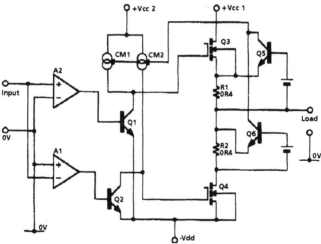

Figure 9.6
Van de Gevel's circuit

The Quad Current Dumping Amplifier

The circuit layout of this amplifier is shown, in its essentials, in Figure 9.7, and its intention is to allow the output transistors (Q1,Q2) to be operated at zero bias and zero quiescent current – a condition which requires no initial setting-up adjustment. Its operation can be considered to lie in either one or other of two distinct modes, that when the output transistors are non-conducting, and that when the output transistors are in their normal operating condition.

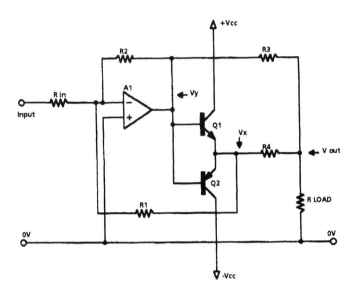

Figure 9.7
Current Dumping circuit

Taking the first of these first, a low distortion small power amplifier, A1, is arranged to drive the load through R3. Since there is in this operating mode no contribution from the output transistors, the signal returned to the input of A1, through the main feedback loop $(R1/R_{in})$, will act to hold the output signal (V_{out}) to the level determined by the feedback signals through $(R1 + R4)/R_{in}$, and $R2/R_{in}$, and, provided that the required load currents are within the capability of A1, the overall system will operate with very low distortion.

However, if the signal output level at V_y is increased to the extent that one or other of the output transistors is driven into conduction, the voltage at V_x, and hence the current through R4 into the load, will be increased, thereby increasing the amount of NFB through R1 to the input of A1, and reducing the gain of A1 to compensate for the added gain of the output transistor which is now in circuit. If R4, R2 = R1, R3 then the extent to which the gain increases or is reduced, as the output transistors pass into or out of conduction, will be exactly compensated by changes in the amount of overall NFB which is applied, and there will be no crossover distortion introduced by the unbiased output devices.

So far, so good. The arrangement proposed by Quad provides a simple and readily understood means of using unbiased output transistors without a major penalty in respect of the system distortion. Unfortunately, the presence of R4 in the output signal path to the load would lead to an undesirable waste of output power, so Quad replaced R4 with a 3.3µH inductor (Z4) and preserved the equivalence of R1, R3 = Z4, Z2 by replacing R2 with a 120pF capacitor (Z2). Though this exchange preserves the intrinsic philosophy of the design it made the actual operation of the amplifier significantly more complex, which allowed those of the Hi-Fi press who resented Quad's aloof approach to outside reviews a pseudo-technical reason for denigrating the performance of this remarkable design.

The Sandman Class S System

As is shown above, in Table 9.1, all of the simple push-pull output layouts shown in Figure 9.3 exhibit a conspicuous reduction in their levels of distortion when their normally low impedance output load is replaced by a high impedance one: a reduction in distortion from 1–2% down to less than 0.02% being typical. This effect was noted by Sandman, who proposed the circuit shown in Figure 9.8 (Sandman, A.M., *Wireless World*, September 1982, pp. 38–39). The way this circuit works is that a low power, very linear amplifier, A1, is arranged to drive the load through R4, while a high power, low quality amplifier is used to drive the bridge network R3–R6 at point B, and, through this, the output load. The voltage comparator, A2, monitors the extent of the bridge unbalance, and drives the output transistors so that the current fed into the resistor bridge at point B ensures that the difference in voltage level between points A and C is very small. By making the signal voltage developed across R4 very small, the high power devices effectively present the small power amplifier with an infinite impedance load. Some crossover effects will, of course, arise during the period in which the base voltage applied to Q1 and Q2 is swept across their non-conducting regions. During this brief excursion A1 will supply the load directly through R4. The resistor values used in the bridge should be low enough to allow this.

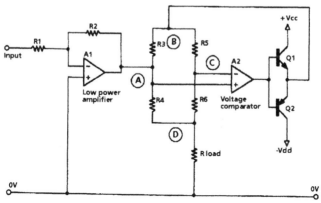

Figure 9.8
The Sandman circuit

The Technics Class AA System

This arrangement is shown in Figure 9.9, and is used in the Technics SE–A1000 power amplifier and all their current power amplifier designs. This circuit has obvious similarities in its approach to the Sandman design, and clearly uses the same basic philosophy in which an output push-pull power amplifier (they call this a class AA stage) operates to provide a very high impedance load on the voltage amplifier. The Sandman/Technics approach offers a unique and valuable way of reducing overall amplifier distortion without requiring high loop gains in the power amplifier, or thereby reducing the loop stability margins.

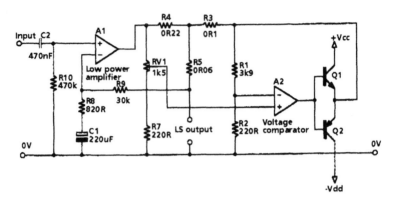

Figure 9.9
The Technics circuit

Voltage Amplifying Stage Gain and Overall THD

In normal amplifying systems employing NFB the open loop gain (i.e. before NFB is applied) will be made as high as feasible – so far as this is possible without introducing an unacceptable level of loop phase error – so that there will be enough gain left over to allow a useful amount of NFB to be applied. The relationship between distortion and negative feedback is:

$$D^{\cdot} \approx D^o/(1+\beta A^o)$$

where D^o is the distortion before NFB is applied, $D^{\cdot}$ is the harmonic distortion after the application of negative feedback, β is the feedback factor (by which I mean the proportion of the output signal which is fed back to the input) and A^o is the open loop gain of the amplifier. This is very similar to the relationship between closed loop gain and negative feedback:

$$A^{\cdot} = A^o/(1 + \beta A^o)$$

where A^o and $A^{\cdot}$ are the respective values of the stage gain before and after the application of NFB. The implication of these two equations is that the distortion is reduced in direct proportion to the reduction is stage gain. However, the relationship is only an approximate one since the presumption is made, in the case of harmonic distortion, for example, that the amplifier gain (and the feedback factor) remain constant as a function of frequency from that of the fundamental to that of the nth harmonic – an assumption which is unlikely to be true, and which is the major reason why the distortion of an amplifier with NFB will worsen as the frequency is increased. It also makes the assumption that the feedback remains negative throughout the frequency range of interest, which may be only marginally correct.

However, if we assume for the purposes of this argument that the distortion is reduced in the same proportion as the gain, and that the closed loop gain is the product of the open loop gain and the feedback factor – which will be nearly true if the gain (A^o) is very high – then for an amplifier with an open loop gain of 1000˘, to which NFB has been applied to reduce the gain to 10˘, if the initial distortion (D) was 2%, then the distortion with NFB applied (D˙) will be 0.02%, and this is the basis of the design for almost all the commercial audio amplifiers on sale in Hi-Fi shops, for which the two performance factors of greatest interest to the (probably naive) purchaser are the total output power and the total harmonic distortion. So far as the THD is concerned this is mainly determined by the amount of NFB which can be applied, which, in turn, depends on the open loop gain which can be obtained from the voltage amplifier stage. This is a function of the circuit structure which is used, and I have shown how the gain can be increased by refinements in the circuit layout in the various diagrams of Figure 9.10.

Basic Gain Stage Constructions

All of the gain stages I have shown are based on the long-tailed pair circuit configuration – sometimes called a differential pair – because this allows the signal to be operated upon between a symmetrical pair of supply lines, and permits the design of amplifiers which do not need a large value output DC blocking capacitor – a component disliked by the Hi-Fi afficionados because of its supposedly variable and voltage dependent characteristics – and is therefore the approach used in all audio amplifier designs having any pretensions to high audio quality. The test conditions for all the circuits were: frequency 1kHz, supply voltage ±15V, 3V rms output.

In the layout of Figure 9.10a, 100% DC negative feedback is applied through R4 and C2 to stabilise the output DC potential, which is held to 0V ±100mV, but, because C2 shunts the AC component of the feedback signal, allows virtually the full AC stage gain (A^o) to be achieved. The current through both Q1 and Q2 is about 2mA. This relatively high value is commonly chosen to increase the slew rate of Q3, between whose collector and base a dominant lag HF stabilising capacitor of 50–100pF (not shown) will normally be connected. Because of the need to ensure similar working conditions for both of the two input transistors R2 must be chosen to provide a voltage drop of about 0.6V at a current of 2mA, which requires an inconveniently low value for this component from the point of view of stage gain, which, in the circuit shown, is 4412˘.

Elaborating the circuit so that Q3 becomes a Darlington pair, Q3/Q4, as in Figure 9.11b, allows R3 to be increased to 680R, and this, predictably, very nearly doubles the stage gain to 9091˘. Since R3 is still a low value, the benefit offered by the relatively high input impedance of the Darlington pair cannot be utilised. If the need to keep the slewing rate of Q3/Q4 as high as possible can be ignored – perhaps by the use of a different method of HF loop stabilisation – the input transistor currents can be decreased and R3 increased to 12k, which will increase the stage gain to 11, 540˘.

A much more profitable approach would be to replace the load resistor of Q1 with a constant current source which would offer a dynamic impedance at least ten times greater than a 680 ohm load resistor, giving a stage gain increase to some 100, 000˘.

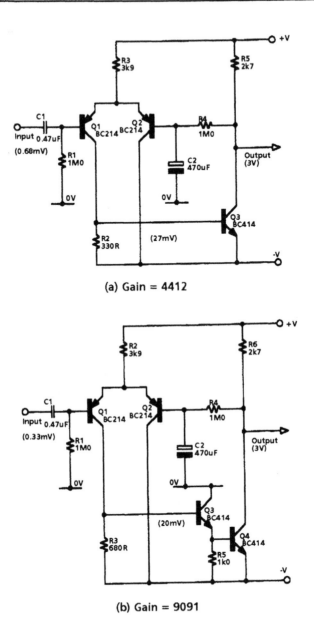

(a) Gain = 4412

(b) Gain = 9091

Figure 9.10a/b
Simple gain stages

Even better, and using no more components than a typical constant current source, would be the use of a current mirror, as shown in Figure 9.10d, which combines the outputs of Q1 and Q2 and will double the gain given by a simple constant current source, to provide a measured stage gain of 187, 000⁻. Ideally a matched pair of transistors should be used for Q5 and Q6 (or, indeed one of the proprietary IC style current-mirror devices, available in various input/output current ratios). Failing this, a low value preset potentiometer should be inserted in the mirror transistors' emitter leads, as shown in Figure 9.10e, to allow the currents through the two input transistors to be balanced.

A practical snag with all of the circuits shown so far is that they offer very little rejection of any intruding signal voltages present on the +ve supply rail, and this

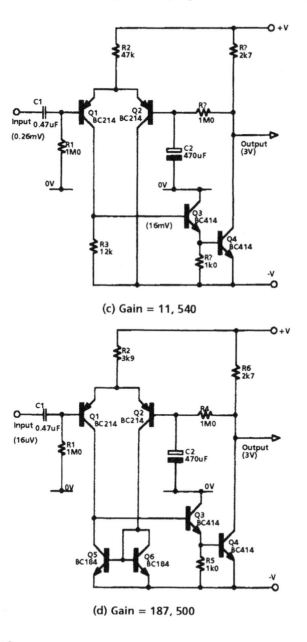

(c) Gain = 11, 540

(d) Gain = 187, 500

Figure 9.10c/d
Simple gain stages

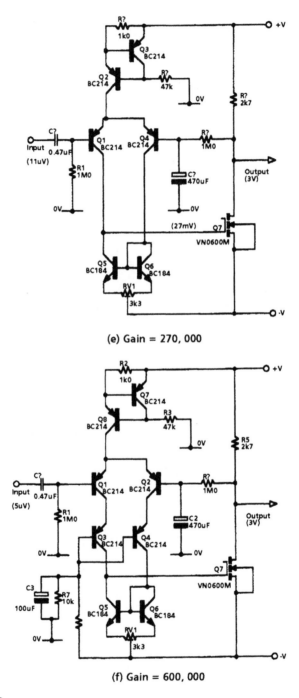

(e) Gain = 270, 000

(f) Gain = 600, 000

Figure 9.10e/f
Simple gain stages

would inevitably degrade the circuit performance. Substituting a two transistor constant current source for R2 in the earlier designs would avoid this problem, and possibly also allow a small further increment in stage gain. In the circuit diagram of Figure 9.10e I have also explored the use of a small-signal T-MOSFET, which is very linear in use as a class A gain stage, and also has a very high input impedance. This increases the gain still further to 270, 000⁻.

A final beneficial circuit refinement, shown in Figure 9.10f, is to insert a pair of cascode connected transistors (Q3 and Q4) in the collector circuits of the input transistors. This has the effect of clamping the collectors of Q1 and Q2 to a fixed DC voltage – approximately 4.5V – so that there is no effective negative signal feedback through the internal resistance and capacitance between collector and base of Q1 and Q2, and avoids any loss of gain in the input transistors due to this unwanted internal feedback. With this refinement the circuit gain is increased yet further to 600, 000⁻.

In all the circuits I have shown so far I have used a simple 2.7k resistive load for the output, although it would clearly be possible to obtain a considerable increase in unloaded stage gain if the output load resistor were to be replaced by, for example, a high dynamic impedance constant current source. This choice of load is quite deliberate, in that the circuits explored have all been of a kind which could be used in the gain stages of an audio amplifier, where the output stage would be a push-pull pair of Darlington connected emitter-followers or a similar pair of complementary MOSFET source followers, for which, I felt, a 2k7 ohm resistor, with the possible addition of a 2000pF shunt capacitor, would be a more representative type of gain stage load. In any practical audio amplifier, the gain stage would be operated with a high dynamic impedance load in order to increase the extent of supply-line signal rejection, but this would not increase the open loop signal gain because of the loading effects of the output devices.

Output–Input Signal Isolation

In most real-life amplifying devices there will be a signal feedback path from output to input, and in a transistor the capacitative component of this will depend on the junction area and the collector to base voltage. For a typical junction transistor of 1W dissipation and 50V V_c, this feedback capacitance (C_{c-b}) will be will be in the range 5–10pF, while for a comparable small-signal MOSFET, C_{d-g} will be about 30–50pF, and the Miller effect due to this feedback capacitance will reduce the gain at higher signal frequencies. This loss of HF gain can be lessened by cascode connecting the output device, as I have shown in Figure 9.11. This somewhat reduces the 1kHz gain to 330, 000, but allows a gain of 90, 000 at 20kHz, by comparison with the circuit of Figure 9.10f, which has a gain of 600, 000 at 1kHz, but only 30, 000 at 20kHz. This is not an additional problem with most conventionally designed audio amplifiers, since in these a capacitor in the range 50–100pF will be connected between the collector and base of the device to cause a feedback stabilising dominant lag, but if some alternative method of loop stabilisation is to be employed, an output cascode connected buffer could be valuable.

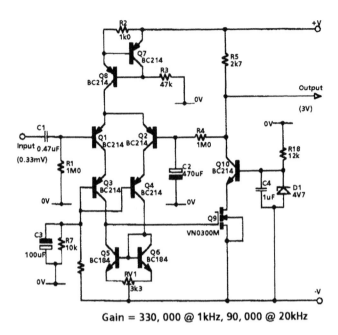

Gain = 330,000 @ 1kHz, 90,000 @ 20kHz

Figure 9.11
Cascode output isolation

Effect of NFB on Signal Distortion

If an open loop gain of 90,000 is available, and the required gain, with feedback, is 30, then a factor of 3000 is available to reduce the distortion of the amplifier. If we assume that the distortion of an output pair of emitter–followers worsens from 2% to 20% as the amplifier output approaches its overload point, then the (theoretical) residual distortion, near overload, could be reduced from 20% to 0.006%. This is a somewhat artificial assumption since there are a number of other factors which will degrade the overall distortion figure, such as the unsuspected generation of signal voltages between points which are notionally at the same potential, or the failure – seen above in some valve amplifier designs – to derive the feedback signal from the point at which the LS load is connected. However, I am left with the feeling that designers who claim THD figures of the order of 0.0001%, or even 0.001%, are talking about the predictions drawn from simplistic computer simulations rather than the actual performance of real pieces of hardware. There are, of course, techniques, such as those suggested by Sandman, by which the distortion of the output stage can be reduced without relying entirely on the improvements brought about by NFB, and these would result in a genuine improvement in results, but these are an exception to the normal practice of commercial audio power amplifier design.

Symmetry in Gain Block Structure

A range of well-known solid-state gain stage layouts has been shown above and in previous chapters, and it is clear that the contemporary audio amplifier circuit designer – unlike his predecessors in the days of valve circuitry – has a very wide choice of component arrangements at his disposal. So far as there is any general trend in contemporary audio amplifier design it is towards the completely symmetrical type of layout such as that due to Borbeley and shown in Figure 9.12 (Borbeley, E., *Wireless World*, March 1983, pp. 69–75). The basic structure of each half of this circuit is similar to that of Figure 9.10c with the added refinements of an output cascode buffer between Q4 collector and the output load, and with a constant current source in place of R2. Because the system is fully symmetrical, and employs active loads (Q10/Q11) the 1kHz stage gain, at 30, 000⁻, is twice that which would be given by the equivalent single-ended layout of Figure 9.10c.

The basic aim of this circuit structure is to lessen or eliminate the inevitable effects of slew rate limiting which will occur under limiting conditions in any single-ended system. This purpose is assisted by the stabilisation of the feedback system by the use of a pair of HF step networks (C2/R10, C1/R6) which do not begin to limit the HF gain until 72kHz, which is well outside the 22kHz design bandwidth determined by R18 and C4. My own judgement is that this type of layout is much superior to the dominant lag stabilised designs of the type shown in Figures 9.10c or 9.10d, which typify so many of the middle-Fi offerings of the present day. This view is obviously shared by a number of major Japanese audio equipment manufacturers who have adopted the essential features of Borbeley's circuit design.

Negative Feedback and Sound Quality

Although it is possible, in principle, to design an amplifying module which exhibits progressively smaller amounts of steady-state harmonic distortion, by the use of negative feedback connected around a gain block of increasingly high gain, there is no guarantee that the amplifier module so made will be satisfactory in use in any given audio application. I will explain.

I feel, sometimes, that it is a pity that audio amplifying and sound reproducing equipment succeeds or fails depending on the reactions it evokes in the ears of those who listen to it. My misgivings arise because the characteristics of the human ear vary from person to person, and from day to day, and since what is heard by the listener is subject to his personal dislikes or preferences, the acoustic performance of the equipment cannot be completely specified, with certainty, by any set of electrical or engineering specifications. One can make guesses that certain features of the design will be a good thing, or that certain other features which one has tried to avoid or minimise would have been bad things, but one can never do better than feel confident that what one proposes or has made will be favourably judged by one's peers.

It is because of this that those electronics engineers who have designed audio amplifiers and other signal handling equipment, which has been judged by their users to perform to their entire satisfaction, enjoy a higher seat in the electronic engineering Pantheon – because of some supposed mastery of the black arts of audio – than other equally competent engineers who had designed, for example, instruments to tell a submarine captain how far he is beneath the surface of the sea, or who has designed, say, a simple, efficient and reliable mechanism for opening and shutting the hatches which enclose the landing gear on an aircraft.

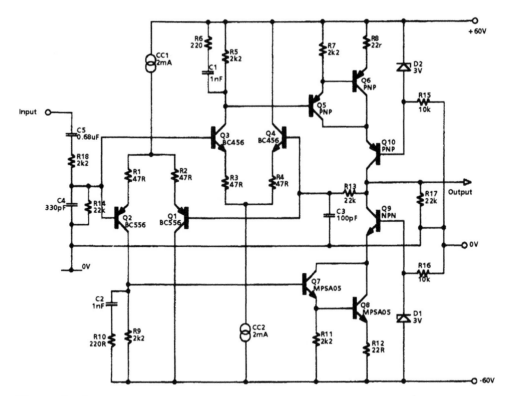

Figure 9.12
Borbeley symmetrical driver

Certain aspects of audio design can be specified with relative ease – such as the power bandwidth, the uniformity of the frequency response over the audio band between, say, 10Hz and 20kHz, the output power which can be delivered to certain specified load impedances at certain levels of distortion, the size of the input signal required to produce the specified output power, and the signal to noise ratio of the equipment or the signal breakthrough from one channel to another, in a stereo or multi-channel system – however, all of these measurements relate to purely steady state characteristics, such as could be made with an input signal derived from a low distortion, constant amplitude, variable frequency sine wave oscillator, and bear only a fleeting resemblance to the nature of the audio signals which are likely to be presented to the equipment. These will consist of a multiplicity of simultaneous signals, each having widely variable amplitudes and rates of change of amplitude, and most of these waveforms will be non-repetitive quite unsymmetrical.

The task of designing a signal source which would simulate all of the relevant aspects of an audio signal in demonstrating equipment shortcomings is a daunting one, but quite a lot of information – particularly that relating to poor loop stability in amplifiers using NFB – can be shown by the use of an input square-wave signal, as I have illustrated in Figures 7.12a to 7.12g, in my book *Audio Electronics* (Butterworth-

Heinemann, 1995, pp. 280–284). Much more needs to be known and specified in relation to any audio amplifier before it can be declared to be satisfactory, but, at the very least, no statement of performance can be regarded as complete without the illustration of the performance, as demonstrated by real-life hardware, under test on reactive loads, photographed on the screen of a wide-bandwidth oscilloscope. This is relevant because this is the area where the inadequacy of HF feedback loop compensation methods will show up – both on the oscilloscope screen and in the ear of the listener. As a supplement to the oscilloscope waveforms, it would be valuable to specify the output square-wave settling time of the system, for various test frequencies, amplitudes and load characteristics, defined as the time required, in microseconds, for the output waveform to settle within ±1% of the required output level, following an input step-function-type change of input signal level, as shown in Figure 9.13.

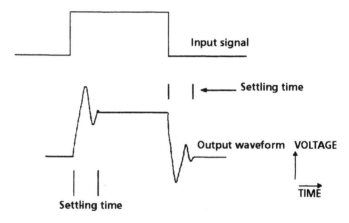

Figure 9.13
Amplifier transient response and settling time

Power Amplifier Output Stages

Although for a time it seemed that the undoubted advantages offered by power MOSFETs as output devices – greatly extended HF gain, freedom from secondary breakdown (allowing simpler methods for output transistor protection), greater intrinsic stability of the DC working point, greater ease of paralleling output devices to allow increased output power, as shown, for example, in the output layout of the Borbeley design, illustrated in Figure 9.14 – would ensure a growing number of MOSFET based audio amplifier designs, the major mass-market manufacturers continue to use bipolar power transistor layouts of the general form shown in the Marantz PM–16 design of Figure 9.15. In this circuit, the forward bias for the required output transistor quiescent current is generated by a simple amplified diode layout built around a monolithic Darlington transistor pair (Q7/Q8), the output transistor overload protection is provided by the simple Zener diode chain across the signal input to Q1 and Q6, and output LS protection is given by a relay cut-out whose contacts (not shown) are situated in the LS output line.

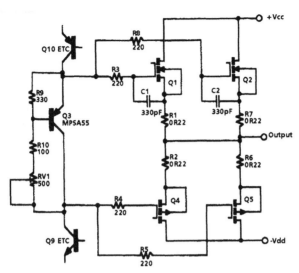

Figure 9.14
Borbeley output layout

The most notable feature of this power amplifier design (whose output is rated from 90 to 150W per channel, depending on the output load) is the elaboration of the normal output complementary Darlington pair into a symmetrical three-transistor cascade (Q1/Q2/Q3 and Q6/Q5/Q4). The cross-coupling of the driver transistor emitter circuits, via R6 and R3/C1, is now quite conventional practice for the driver stages of power junction transistors since the reverse biasing of the output transistor base circuits helps drain the stored holes from the base-emitter region and thereby speeds up the turn-off time of the transistor following any period in which it has been driven momentarily into saturation.

A virtually identical output configuration – following a Borbeley-style symmetrical driver stage – is used by Rotel in their RHB10 200–330 watt power amplifier, though in this, the output transistors Q3 and Q4 in Figure 9.15 are replaced by four parallel connected devices as each half of the output emitter-follower pair, current sharing being ensured by the inclusion of a separate 0.22Ω emitter resistor and a 10Ω base resistor for each transistor. Once again, the input bias needed to establish the output quiescent current is achieved by a simple amplified diode, and output LS protection is afforded by a relay cut-out whose normally closed contacts are located in the LS circuit.

Output DC Offset Control

Almost all contemporary audio amplifiers are based on the split-supply rail philosophy, in which the input–output signal path operates about a zero mid-point voltage, and the output of the amplifier is connected by a DC path to the LS output terminals. This type of circuit layout demands that there shall be no significant (say, not greater than ± 10mV) DC offset present across the output terminals, and this is now achieved, almost universally, by the use of an op amp in the input signal path as shown

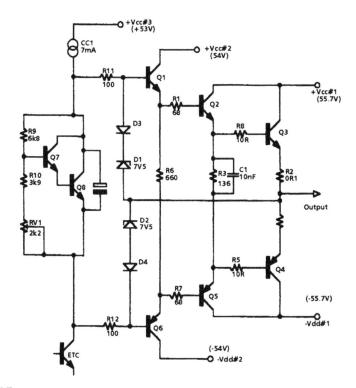

Figure 9.15
Marantz PM–16 output layout

in Figure 9.16, since most off-the-shelf op amps will have an intrinsic DC offset within a few mV, even without any adjustment. Unfortunately, this type of offset removal, though effective and inexpensive, brings with it the problem of switch-on plop, since it is certain that the voltage required at the inverting input of A1, for zero DC offset at the power amplifier output, will not be zero, and until the capacitor C1 has charged to the required voltage through R4 there will be an unwanted amplifier output voltage. The longer the time constant of R4 and C1, the longer this error voltage across the LS load will persist, and the time constant cannot be shortened without impairing the amplifier LF gain.

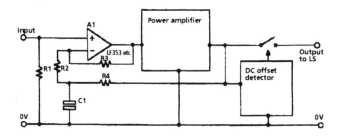

Figure 9.16
DC output offset control

If there are relay operated switching contacts in the LS line, as shown in Figure 9.16, used to protect the LS load from any circuit failure which would generate a damaging DC offset at the LS output terminals, then this relay circuit can be used to isolate the LS from the amplifier output until there is zero residual DC present at this point following switch-on. The effectiveness and simplicity of this method has resulted in its widespread adoption among commercial systems. However, these switch contacts may have to handle high output currents in use, so very good quality switching contacts are essential if signal degradation is to be avoided.

Hybrid Systems

A number of hybrid systems have been proposed, such as that by Mortensen, using valve driver stages and MOSFET outputs (Mortensen, H.J., *Audio Amateur*, January 1989, pp. 14–20) or Macaulay, using transistor driver stages and pentode valve outputs (Macaulay, J., *Electronics + Wireless World*, October 1995, pp. 856–859), intended either to take advantage of the robustness of transformer-coupled valves in respect of overload, or the freedom from the need to use an output transformer offered by low output impedance solid state designs, but these ideas seem to have had negligible impact on mainstream circuit design which, apart from what I consider to be the hunting grounds of wealthy eccentrics, has remained unshakably solid state.

Listening Trials

The purpose of Hi-Fi magazines is to sell Hi-Fi magazines, and, in the process, to make money for their publishers, to give employment to their staff and to provide entertainment and instruction for their readers. In this respect they are no different from any other periodical, with the possible exception that their contributors must deal, to a significant extent, in informed opinion rather than in simple statements of fact, and this aspect of their editorial content comes very much to the fore in their judgements of audio amplifiers, since the basic specifications of so many of these are so similar that a judgement by the potential purchaser, solely on the basis of specified performance, would be very difficult. Because of this, there has been a growth in the numbers of reviewers who have made a speciality of the assessment of the quality of the sound produced by any given audio power amplifier (or, to a lesser extent, preamplifier) when it is used to reproduce any particular piece of music.

Inevitably, personal preferences on the part of the reviewer, in respect of the styling or convenience of the equipment, or the general helpfulness or generosity of the manufacturers, must, with the best will in the world, spill over into the area of the reviewer's technical assessment of the performance of the unit, and the reviewer's judgement of its tonal quality. This may not matter very much to the average reader, who is probably unlikely to want to buy that particular product anyway, and may quite enjoy reading slanging matches between august names, but it is a matter for considerable concern for the manufacturers who may feel that their products are being unfairly denigrated by a coterie of prejudiced or hostile reviewers, or by engineers who feel that the wrong lines of technical development are being extolled. For this reason, both the manufacturers (to defend their products from unfair attack) and the magazines and their reviewers (to establish in the eyes of their readership their ability to make valid subjective judgements) have, from time to time, set up comparative performance trials in which panels of listeners have been invited to record their preferences for one or other amplifier or other piece of equipment, which should, ideally, be randomly selected by a third party – or, indeed, by some electronic sampling mechanism without any human interest in the proceedings or their outcome.

An early comparative trial along these lines was commissioned by Quad, and reported by Moir (Moir, J., *Wireless World*, July 1978, pp. 55–58). In this trial, individual members of a panel of six skilled jurors, mainly drawn from those having an engineering involvement in sound reproduction, were asked to express their individual preferences for one or other of a pair of units presented anonymously in randomly ordered sequential tests. These tests could be between different amplifiers or (as a check on the panel's ability to distinguish similarities) between the same unit twice repeated. The units compared were Quad II valve amplifiers, Quad 303 transistor amplifiers, and the Quad 404 Current Dumping power amplifier – a unit which does not seem to have been fairly treated by the young cognoscenti of the Hi-Fi press. When the results of this carefully prepared and executed trial, which involved some 373 individual tests, were subjected to statistical analysis the results indicated that, in the judgement of the panel, there was no obvious and significant preference for one unit rather than another, and that the members of this highly skilled listening panel were generally unable to distinguish one unit from another.

The results of the Quad trial were so clearly at variance with the long-term experience and expectations of the Hi-Fi press or their contributing reviewers that a repeat trial was organised by Martin Colloms and *Hi-Fi News* (HFN/RR November 1978, pp. 110–117), and reported on by Colloms and Adrian Hope. The purpose of this repeat trial, with a initial judging panel of seven members, largely drawn from the Hi-Fi press, was to try to eliminate the possible factors which might have blurred the distinctions between different systems, a confusion of outcome which the critics felt must have occurred in the trials sponsored by Quad. This second trial employed a Quad 405, a Naim NAP 250 and a TVA Export (high power valve operated) amplifier – and so embraced a wide range of technologies as well as a wide spectrum of popular esteem. In the event, the results were remarkably similar, in that, with two exceptions, the (again highly experienced) panel was unable to differentiate between the units under test.

Since great care and self-control had been exercised by the members of this panel to avoid influencing each other's judgements it was surprising to discover that there was obvious evidence of cross-influence in the results. However, ignoring this discovery, which did not, in this case, affect the results, the outcome was largely the same as that from the Quad trial, that the panelists were, as a whole, unable to decide which unit they were listening to, or whether any one unit was superior to another. Once again, it was felt by Colloms that the outcome was, in some way, distorted, perhaps by the very large number of musical excerpts to which the panel had been asked to listen or the relatively short length of each trial, so the tests were repeated, with a larger (13 member) team of jurors, who were asked to adjudicate on their own, and were allowed to increase the duration of each trial as wished. Once again there was no clear evidence that there were audible differences between the units under test, apart from the suggestion that two of the panelists were able to identify individual amplifiers, perhaps because of small residual differences in their frequency response.

Speaking for myself, as an amplifier designer who listens to the results given by his designs, I know quite well that there are audible effects which result from various identifiable and measurable electrical design defects, and measurements on those commercial designs to which I have had access have frequently revealed the presence of uncorrected electrical defects of kinds which I have learnt to associate with certain faults in sound quality. However, even when one has trained one's ears to recognise these faults, this recognition may take a little while.

At this point, may I introduce Fingal, my Siamese cat, who had a great liking for Bach organ music, and would climb onto my lap, and purr contentedly whenever I played one of his favourite pieces from gramophone record. If I had made any change in the reproduction system he would be aware of this, and signify his awareness by the angle or alertness of his ears. If, after a minute or two, he was prepared to accept the change he would settle down to (apparently) enjoy the music. On one occasion I was lent, for lecture purposes, an expensive and impeccably specified Hi-Fi amplifier. As soon as I began to play music through this new system Fingal leapt from my lap and fled from the room. Actually, I agreed with his judgement.

My conclusion, clearly shared by Fingal, is that there are audible differences between amplifiers, but, between good units, the differences are small, and not always easy to identify in a short listening session. I also believe that most of the more experienced Hi-Fi reviewers are sensitive to these differences, but tend greatly to exaggerate their importance. In my experience there are greater differences between the sound of comparable LS units, or phono cartridges, or even CD players, mainly for clear and identifiable and measurable reasons. As engineers become better at identifying and minimising measurable faults, then also these differences will diminish.

CHAPTER 10

PREAMPLIFIERS

The function of a preamplifier is to allow the user to select a signal from one or other of a variety of sources – such as a radio, a tape recorder, a gramophone pick-up or a CD player – to offer an input impedance and signal level handling capacity which is appropriate to the chosen source, to adjust the amplitude/frequency characteristics of the input signal, if necessary, to give the required tonal quality, to amplify the signal, if required, to the 0.5–5V rms level needed to drive an audio power amplifier, and, finally, to provide this output signal at an adequately low impedance level.

In earlier, thermionic valve operated audio systems the minimal functions of a preamplifier were often performed by a simple selector switch and volume control at the front end of the power amplifier, or perhaps by an additional valve, with facilities to switch between one or two alternative frequency response shaping networks. These would be used to provide selective attenuation or gain adjustments, either by passive RC networks or by negative feedback arrangements.

It was unusual at this period to employ a separate preamplifier unit, so where more extensive signal manipulation circuitry was provided – as in the case of the Williamson, or the Brimar preamplifier shown in Figure 5.16 – it was usually incorporated within the main amplifier housing, an arrangement which is now described as an integrated amplifier. It was not until the advent of the transistor, with its small size, its low power requirements, its absence of AC operated heater circuitry and its low heat generation, that the advantages of segregating the low signal level input circuitry and the high signal level power stages and power supply wiring could be seen as a normal feature in high quality systems.

It is also required that the signal manipulation and amplification carried out by the preamplifier shall be done without significantly impairing the distortion or the signal/noise ratio of the input signal, though the weight attached to the term significantly will probably be a matter for debate between the designer and his critics – who may be insulated by their armchairs from the need, in practice, to make accommodations between desirable, but not always mutually compatible, requirements.

A friend of mine, for whom the purchase, use, and subsequent resale of Hi-Fi exotica appeared to be a major hobby interest, told me that in his experience there were greater differences in the sound of different preamplifiers than there were between comparably well-designed power amplifiers. If he was correct in this belief, it is probable that the tonal differences he had noted had their origins in the various parts of the circuit in which the gain/frequency response of the system had been deliberately modified, for example to compensate for the skewed amplitude/frequency response used in the recording of an LP or EP gramophone disc.

Gramophone Record Replay Equalisation Characteristics

Various combinations of recording pre-emphasis and replay de-emphasis have been proposed for use in gramophone record reproduction. Of these, the most relevant is the method recommended in the RIAA:BS1928/1965 specification, in that this has been adopted, on a world-wide basis, for the manufacture of 33 and 45rpm vinyl discs.

The proposed pre-emphasis characteristics were based on a recognition of the practical problems in disc manufacture and reproduction using velocity sensitive (e.g. electromagnetic) cutting and replay heads. These realities also influence the extent to which electrical headroom is needed – a matter which I will explore later on. In its original form, the RIAA specification required a replay characteristic of the form shown in Figure 10.1, in which the shape of the replay response was defined by three time constants, 3180µs, 318µs and 75µs. There has, however, been a recent amendment to this specification, in acknowledgement of the inevitable VLF noise present on LP disc replay, to include an additional LF roll-off defined by a 7950µs time constant, as shown in the dashed line curve in Figure 10.1.

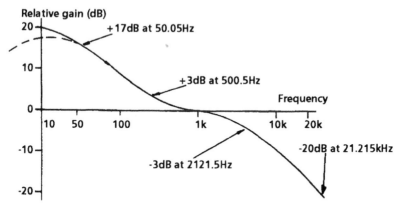

Figure 10.1
RIAA response curve

The need for the rising replay gain characteristic below 1kHz is because of the typical maximum allowable LP/EP groove separation of 0.01cm, which limits the magnitude of the permissible recording cutter excursion (to avoid adjacent groove breakthrough) with a consequent linear reduction in replay stylus velocity (and output voltage from the pick-up) with decreasing frequency. The concept of HF recording pre-emphasis, followed by replay de-emphasis, was initially adopted for 78rpm records, so that the reduced level of treble, on replay, would lessen the irritating audible hiss due to the emery powder loading of the shellac discs. This arrangement has been retained even in the case of the much quieter vinyl surfaces. The most commonly used of the several 78rpm equalisation specifications was similar to the RIAA but with time constants of 3180µs and 450µs to 1kHz, and then with a fall off in gain defined by a 50µs time constant to 21kHz.

Replay Equalisation Circuitry

There are a number of circuit arrangements which will generate the required RIAA replay response curve for use with velocity sensitive gramophone pick-up transducers, and I have shown the more commonly used circuits in Figure 10.2. Of these, the

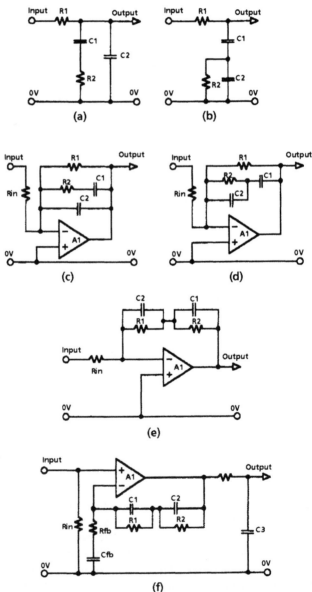

Figure 10.2
RIAA replay circuits

simplest, and therefore, in the opinion of the purists, the most desirable options are the two passive networks shown in Figures 10.2a and 10.2b. However, in order for these to function correctly they must introduce an attenuation of 20dB at 1kHz, and this will require some additional amplification to restore the original signal level. Also, for the frequency response of the networks to conform accurately to the RIAA curve, it is

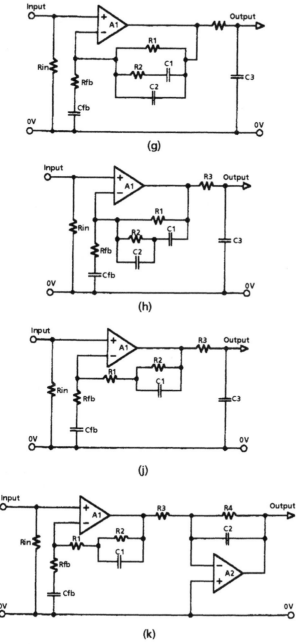

Figure 10.2 continued
RIAA replay circuits

necessary for them to be driven from a very low impedance source and for the output load to offer an impedance which is high in relation to R1. In practice, these requirements imply the need for further amplification or buffer stages preceding or following the passive network, so the presumed advantages of simplicity and absence of (possible distortion introducing) active components are lost.

The layouts of Figures 10.2c, 10.2d and 10.2e can be arranged so that they can be fed directly from the pick-up cartridge, with R_{in} chosen to provide the manufacturers specified loading for the cartridge. This layout also offers a low output impedance to any following stages which avoids the need for additional pre- and post-amplification.

All three of these circuits use shunt feedback, which has the relative disadvantage that the input thermal noise is determined by the source impedance in series with R_{in}, which will be worse, at low to mid-frequencies with an inductive transducer, than that given by a equivalent series feedback layout. A simple calculation would suggest that the source noise level for a series feedback circuit, such as that of Figure 10.2g, when used with a pick-up cartridge having a DC winding resistance of 2200 ohms (connected in parallel with a 47 000 ohms load resistor) will be 0.83µV. (This figure is the room temperature value seen at the input of a feedback circuit of the kind shown in Figure 10.2g, with a 20Hz–20kHz measurement bandwidth.) The calculated value for a shunt feedback system, of the kind shown in Figure 10.2e, with an input (pick-up load) resistor of 47 000 ohms, will have a noise voltage of 4.24µV under the same measurement conditions.

However, these figures are modified by the fact that the measurement bandwidth for an RIAA equalised gain stage is not 20kHz, but less than a tenth of that figure, which changes the noise output figures to 0.24µV and 1.22µV respectively. Also, pick-up cartridges will have a significant value of inductance – the quoted value for a Shure M75 cartridge being 450mH – and this will increase the source noise voltage at the upper end of the audio pass-band. The effect of this is to alter the relative character of the noise given by these two systems, so that the noise from a series-type circuit is a high pitched hiss, while that from a shunt-type circuit is more of a low pitched rustle.

There are also a few other advantages offered by shunt feedback RIAA circuits such as those shown in Figures 10.2c, 10.2d and 10.2e. Of these, the most important, in practice, is that they are largely immune to very short duration input overload effects, which can be caused by voltage spikes in the cartridge output arising from dust and scratches on the surface of the record. These short duration overloads can greatly exaggerate the clicks and plops which are such a nuisance on vinyl disc replay. Also, they have, in theory, a lower level of distortion (see Taylor, E.F., *Wireless World*, April 1973, p. 194) – although this is a matter only of academic interest in the presence of the 0.5–1% THD generated at 1kHz by a typical pick-up cartridge (Linsley Hood, J.L., *Hi-Fi News and Record Review*, October 1982, pp. 59–68).

Shunt feedback layouts can also give a precise match to the RIAA specification, which a series feedback circuit, as it stands, will not. This is because any series feedback layout having a feedback limb which decreases in impedance with frequency will tend to unity gain at high frequencies, whereas the RIAA specification stipulates a gain curve which tends to zero. Unfortunately, in many preamplifiers of commercial origin, which use series feedback RIAA stages in order to obtain the lowest practical (input short-circuited) noise levels, this error is ignored, even though its effect can easily be seen on an oscilloscope (Linsley Hood, J.L., *Wireless World*, October 1982, pp. 32–36) and the tonal difference which it introduces can equally easily be heard by the listener.

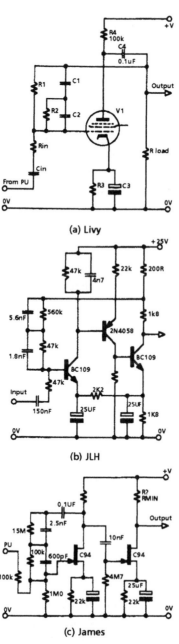

(a) Livy

(b) JLH

(c) James

Figure 10.3
RIAA input stages

This error has been corrected, in the case of the circuit layouts shown in Figures 10.2f, 10.2g and 10.2h, by adding a final RC lag network, R3/C3, in which the required component values will depend on the gain of A1 at 20kHz. The gain reduction in the 1kHz–21kHz part of the pass-band required by the RIAA spec. can be made more simply by dividing the circuitry used to provide this response into two parts – as in Figure 10.2j – where R1, R2 and C1 generate the frequency response curve from 30Hz to 1kHz and R3 and C3 provide the 75µs time constant needed for the 1kHz–21kHz part of the spectrum. A possible drawback with this circuit (which will also apply to any final frequency response correction lag network), depending on the nature of the following circuitry, is that the time constant of R3/C3 will be affected by the circuit load impedance, which is, for the purposes of this calculation, in parallel with R3. This difficulty is removed by the use of an active output integrator, shown as A2 in Figure 10.2k, where R4/C2 provide a 75µs time constant. The only likely problem with this arrangement is that the circuit introduces a phase reversal, which may not be desired.

The required component values for these networks were originally calculated by Livy and quoted in an article describing a simple RIAA equalising stage (Livy, W.H., *Wireless World*, January 1957, p. 29) shown in Figure 10.3a. The calculations were recast by Baxandall in a slightly more accessible form, and quoted in a reference work (Baxandall, P.J., *Radio, TV and Audio Reference Book*, S.W. Amos, ed., Newnes-Butterworth, Chapter 14), in the course of an excellent review of RIAA equalisation options.

Using the formulae quoted by Baxandall, the required values for R1,R2,C1 and C2 in Figures 10.2a, 10.2c and 10.2g can be determined by:

R1/R2 = 6.818
C1 R1 = 2187µs
C2 R2 = 109µs

For Figures 10.2b, 10.2d and 10.2h the component values can be obtained from:

R1/R2 = 12.38
C1 R1 = 2937µs
C2 R2 = 81.1µs

Similarly, in Figures 10.2e and 10.2f, the values are given by:

R1/R2 = 12.3
C1 R1 = 2950µs
C2 R2 = 78µs

Practical Circuit Layouts

In valve amplifiers the most common RIAA equalisation circuitry was of the kind shown in Figure 10.3a, due to Livy, in which the frequency dependent feedback network is connected between anode and grid of a high gain amplifying valve –

typically a pentode of the EF86 type. Translating this type of circuitry into a transistor form using a single bipolar junction transistor (see Tobey, R., and Dinsdale, J., *Wireless World*, December 1961, pp. 621–625, and also Carter, E., and Tharma, P., *Wireless World*, August 1963, pp. 376–379), encountered the two problems that the input impedances were much lower than was the case with a valve amplifier and that the available gain was inadequate to achieve the required frequency response from the stage, especially at the low frequency end of the RIAA curve where a gain of +20dB with respect to that at 1kHz was needed. The normal approach to this problem was to increase the value of R1, which increased the LF gain at the expense of accuracy in conforming to the lower end of the RIAA curve.

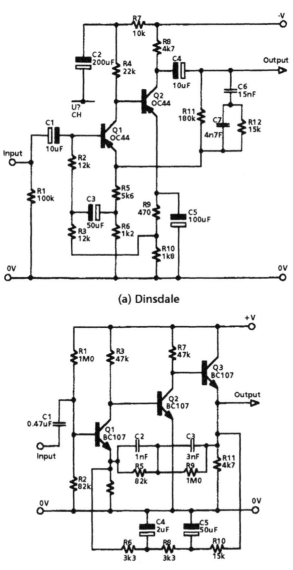

(a) Dinsdale

(b) Bailey

Figure 10.4
Further RIAA stages

In a rather later circuit of my own (Linsley Hood, J.L., *Wireless World*, July 1969, pp. 306–310), shown in Figure 10.3b, I elaborated the single transistor circuit into a triplet, which greatly increased the available gain and also allowed a third order (18dB/octave), 30Hz high pass rumble filter to be implemented using the negative feedback path between the two BC109 emitters.

The availability of junction FETs, which approached much more closely in their operating characteristics to the thermionic valve – with the exception of a much lower gain – encouraged their use in circuits based on their valve precursors, and an FET based design due to James, using some early devices of this type (James, D.B.G., *Wireless World*, April 1968, p. 72), is shown in Figure 10.3c.

The circuit due to Dinsdale (Dinsdale, J., *Wireless World*, January 1965, pp. 2–9), shown in Figure 10.4a, was one of the earliest of the transistor based RIAA equalisation circuits to use series feedback around a gain block having a useful amount of stage gain, and became the standard design adopted by a large number of Hi-Fi manufacturers, though with silicon planar transistors rather than the germanium diffused junction types used by Dinsdale. A problem with this design was that the equalising network applied a capacitative load on the output of Q2, which reduced the gain of Q2 at the upper end of the audio spectrum, although, in practice, this might go some way to remedying the inadequate HF attenuation which is characteristic of the series feedback RIAA system. This problem of falling HF gain was remedied by Bailey (Bailey, A.R., *Wireless World*, December 1966, pp. 598–602) by the use of a transistor based version of the valve Ring of Three shown in Figure 10.4b. This had a low impedance, emitter-follower output stage, and also incorporated a second-order high pass rumble filter, via R6–R10, to attenuate signals below 30Hz.

With the growing use of symmetrical +/– power supply lines, the discrete component version of the RIAA stage shown in Figure 10.5 has become popular, either in the form shown, or in its complementary version, and usually with some elaboration of the

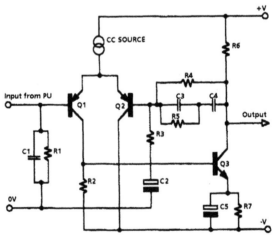

Figure 10.5
Direct coupled RIAA stage

general structure. For a simple design with normal bipolar junction transistor inputs, PNP input devices will normally be chosen because of their better noise performance, although this design decision will be influenced by the availability of application specific components such as the NPN super match pairs described below. This circuit arrangement illustrates one aspect of modern design practice, which favours DC coupling to minimise the use of capacitors in the signal path (except when their use is forced, as, for example, in the RIAA frequency correction network). If Q1 and Q2 were a matched pair it would be possible also to dispense with the DC blocking capacitor (C2) in the feedback line. The capacitor, C1, across the pick-up load resistor will have a value recommended by the cartridge manufacturers to obtain the flattest possible frequency response.

Most contemporary RIAA equalising circuits are based on versions of Figures 10.2g or 10.2h, though there is an increasing trend to the use of versions of Figures 10.2c and 10.2d in which the output from the pick-up cartridge is amplified by a low noise, flat frequency response gain stage, and then presented to the RIAA stage at a signal and impedance level at which the noise level differences between shunt and series connected NFB networks are insignificant. The differences, at the high quality end of the market, between one manufacturer's products and another, now relate to whether integrated circuit or discrete component gain blocks are used as the amplifying stages, and, if the latter, the type of component layout chosen.

Very Low Noise Input Systems

Towards the end of the vinyl era – to the great regret of many gramophone record users, the manufacture of the 12 inch LP disc was discontinued in the late 1980s by the major record manufacturers in favour of the more robust and less expensive compact disc – the gramophone pick-up cartridge design which had come to be preferred by the connoisseurs was one using a moving coil type of construction, rather than the moving magnet or variable reluctance style of mechanism which had dominated the high quality end of the audio market for the previous forty years. The lower mass of the moving coil transducer, and its more effective linkage to the stylus assembly allowed an improvement in the dynamics and stereo channel separation of the system, and the much lower inductance of the pick-up coils greatly improved its high frequency response. However, the typical mean signal level output voltage from such a cartridge might well be only some 50–500μV rather than the 3–5mV output voltage which would be given by a moving magnet transducer, and this required a different approach to RIAA equalised input stage design.

Moving coil pick-up cartridges had been manufactured for many years before there were suitable components for the construction of electronic head amplifiers, and the necessary amplification was provided by an input step-up transformer of the type which had been employed for the same purpose with low output-voltage microphones. However, the use of an electronic gain block offered advantages in terms of wider bandwidth and greater freedom from unwanted hum pick-up.

The main approach employed by the audio circuit designers was to place this gain block between the pick-up cartridge output and the input to the RIAA stage. This way was chosen because it was easier at that time to design a flat frequency response head amplifier than to make a switched gain RIAA stage with a noise level which was fifty times lower than the current norm. For such a gain block to have such a low input noise level it would need a lower input circuit impedance than would be found in a normal small-signal transistor amplifier. There are various ways by which this could be achieved. Of these, the most obvious was to connect a number of input devices in parallel, as shown in the circuit of Figure 10.6, due to Naim. Here five transistors have been arranged in this way, which would, other things being equal, reduce the input base spreading resistance (that distributed resistance between the base connection on the chip and the active base region of the transistor) to one fifth, and reduce the input noise voltage to $1/\sqrt{5}$ or a little less than a half. Some of this input noise improvement is lost, unfortunately, because of the need to include additional emitter resistors to ensure current sharing between the unmatched input devices.

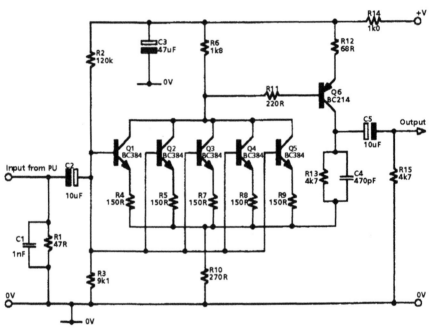

Figure 10.6
Naim NAC 20 head amp

Another approach, rather more suited to the DIY constructor than the large-scale manufacturer, is to use a small-power transistor as the input device, because these will have a larger chip size and, in consequence, a lower base spreading resistance. However, the use of such input transistors might require a degree of selection to find those having the best noise level. An example of this type of circuit is shown in Figure 10.7, for a 30/60˝ head amplifier, particularly suited to very low impedance, low output voltage cartridges such as the Ortofon MC20. In this, Q1, a BD435 small-power transistor, is operated as a grounded base amplifier with a load resistor (R3) which is bootstrapped to increase the stage gain, while NFB is applied via R7, R4 and

R5. The output impedance at Q1 collector is reduced to a low level, as seen at the output, by Q2 and Q3, which are connected as a high gain, compound emitter–follower. The circuit is designed to be powered from a pair of 1.5V AA cells, from which the current drain is 1.5mA/channel. The ±1dB bandwidth is 10Hz–40kHz, and the THD is less than 0.02% at 1V output at 1kHz.

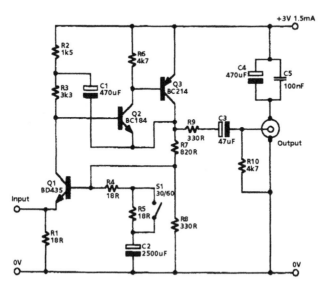

Figure 10.7
Grounded base head amp

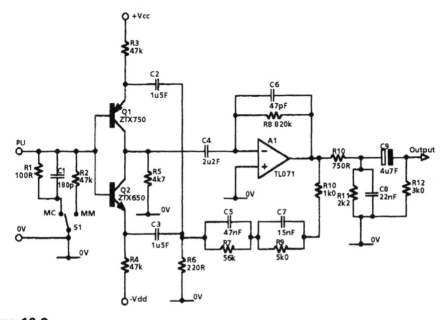

Figure 10.8
Quad MC/MM input circuit

The other way in which the input resistance of the amplifier may be reduced is by operating the input devices in push-pull, which will double the gain and halve the effective noise resistance. An elegant way of doing this is that used by Quad, and shown in Figure 10.8. This takes advantage of the fact that a silicon bipolar junction transistor can operate as an amplifier, though admittedly only at relatively low input signal levels, when its collector voltage is no greater than that on its base. Operation at such a low collector voltage will help to reduce the shot noise component due to collector-base leakage currents in the input devices. Quad used this layout in all their pre-1994 RIAA input stages, with component values being chosen to suit their use either as moving magnet or as moving coil input arrangements. Although the main op amp gain block, A1, is operated in the shunt feedback mode, the circuit as a whole uses series FB because of the interposition of Q1 and Q2. R10, R11 and C8 provide the necessary additional HF attenuation to correct the inevitable series FB gain error.

Low Noise Devices

Bipolar junction transistors

The concept of using multiple, parallel-connected, input transistors as very low noise input devices offered an opportunity to the IC manufacturers to use their existing fabrication techniques to make low noise matched pair transistors where each transistor was, in reality, a large number of parallel-connected devices – distributed across the face of the chip to average out their characteristics. The National Semiconductors LM194/394 devices were early examples of this type of construction, and offered input noise resistance figures of the order of 40 ohms, and bulk (emitter circuit) resistances of about 0.4 ohms. Recently, Precision Monolithics have offered their MAT–01 and MAT–02 super-match pairs, and their SSM–2220 and SSM–2210 equivalents, which are respectively NPN and PNP matched dual transistors, which have input noise resistance values less than 30 ohms. An ultra-low noise gramophone pick–up input head amplifier stage, based on the SSM–2210 and SSM–2220 super-match pairs, is shown in Figure 10.9.

This circuit makes quite an interesting design study, with the Q2 pair of transistors acting as a cascode load for Q1 to increase the input gain, with the Q3 pair acting as a current mirror to sum the outputs of the two halves. Because the bases of Q2 are held at approximately 3V, the collector voltages for Q1 are clamped at about +2.4V, which offers adequate input headroom while still being low enough to minimise collector-base leakage currents. Constant current sources form tail loads for Q1 and collector loads for Q4. In addition to increasing the stage gain, this assists in rejecting signal breakthrough from the power supply rails. Switch S1 allows a choice of input resistances and a choice of stage gain levels, 15 or 150˘, to suit MM and MC pick-up inputs.

Junction FETs

Very low noise JFETs have been available for at least ten years from specialist manufacturers such as Hitachi, Siliconix and Intersil, and, more recently, have also been readily obtainable in dual matched pairs, specifically intended for use in amplifier input stages. The availability of FETs in matched pair forms avoids two of the problems inherent in these devices; that the drain current vs. gate voltage relationship is influenced by so many factors, such as doping level, channel geometry and the distribution of doping impurities, that their drain current cut-off voltage, [$V_{gs(off)}$] and their drain current at zero gate voltage (I_{DSS}) are very variable from one device to another, so that a considerable degree of selection would be necessary to obtain a balanced pair.

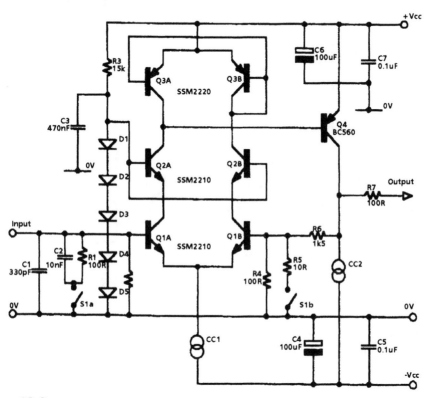

Figure 10.9
Super-match pair head amp

In terms of noise levels, JFETs traditionally suffered from a higher level of flicker (1/f) noise than junction transistors. Also their channel resistance is much higher – because of their method of construction – than those of BJTs. Improvements in 1/f noise have been made by greater care in their physical construction and the channel resistance has been lowered by connecting a large number of channels in parallel, in a similar way to that used in making a junction transistor super-match pair.

A very high performance RIAA corrected input stage, due to Spectral Inc., is shown in Figure 10.10. This also has a cascode connected input stage but with the bias offset voltages of Q2a and Q2b used to provide the 2–3V drain voltage required by Q1a and Q1b. Q3a/Q3b provide a second gain stage, with Q4a and Q4b forming a current mirror collector load for Q3. A symmetrical emitter–follower pair is used to lower the preamplifier output impedance, to allow it to drive a low resistance RIAA equalising network in the feedback path to Q1b. As is current Hi-Fi practice in the USA, aimed at lowering LF phase errors, there are no DC blocking capacitors in the signal path, so RV1 is used to trim out any small residual DC output offset.

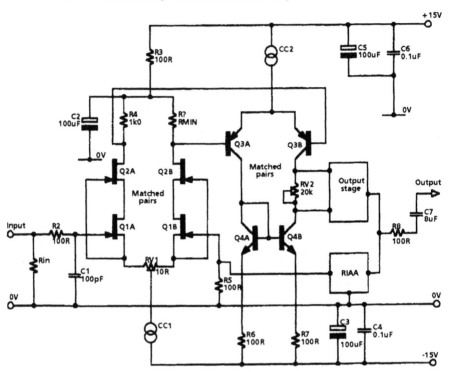

Figure 10.10
Spectral RIAA input stage

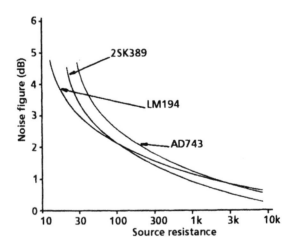

Figure 10.11
JFET/BJT noise figures

The input noise resistance of an FET, at drain voltages which are low enough to avoid significant drain-gate leakage currents, is determined principally by the resistance of the channel, and this is related to the slope (g_{fs}) of the drain current/gate voltage curve, in an ideal device, by the equation:

$$R(n) = 0.67/g_{fs} \text{ (ohms)}$$

As an indication of the relative performance of contemporary JFET and bipolar transistors, the noise characteristics of LM194 and 2SK389 monolithic matched pairs are shown in Figure 10.11. This shows that, for input circuit resistances of above 100 ohms, the 2SK389 FET has a lower noise figure than the best of the bipolar devices.

Low noise op amps

Inevitably, all improvements in device manufacture, as applied to discrete components, will find their way into the field of IC manufacture, and this results, in the present case, in the availability of audio type ICs, such as the Analog Devices AD743 bifet op amp, which has a 10kHz noise voltage of 2.8nV/√Hz and an output distortion figure of 0.0003% at 1kHz and 5V rms. I have also plotted the noise figure of the AD743 on Figure 10.11, on which it appears little worse than that of the 2SK389. Under the circumstances, there seems to be little point in using elaborate discrete component circuits such as those shown in Figures 10.9 and 10.10, as an alternative to an IC, except to provide greater headroom than that would be available from an IC op amp operating from ±15V supply rails.

Headroom

Very few topics can have generated as much debate in the audio field as the extent to which the designer should allow for possible input overload in any given circuit, a factor generally referred to as headroom. Sadly, much of this debate has been entirely misguided. The reason for this is that all the sources which provide signals to the preamplifier do have specific output voltage limits. For example, in the case of a tape recorder, either reel-to-reel or cassette type, the recording level is generally chosen so that the peak output signal is not more than 3dB (√2⁻) greater than the normal span of the recording range. At +6dB, the output THD would probably have increased, in the case of a cassette recorder, from about 0.5% to some 3–5%, and at +12dB (4⁻) magnetic saturation of the tape would probably have clipped the peaks of the signal anyway, preventing any further output voltage increase.

In an FM tuner, constraints on the permissible frequency excursions of the broadcast signal impose stringent limits on the magnitude of the modulation, and this is rigidly peak limited before being broadcast – again giving a +6dB maximum likely signal over-voltage limit. In the case of CD the recording engineer will want to use the bulk of the available 16-bit (65, 536) output levels, while there is an absolute prohibition on any attempt to go beyond this, so, once again, there is an output voltage ceiling, which is unlikely to be greater than +6dB above the intended maximum recorded level.

This leaves only the vinyl gramophone disc as a source of possibly unlimited peak output voltages, but, here again, at least for recorded signals, there are output level constraints, as shown in Figure 10.12. The recording characteristics of the vinyl disc were analysed by Walton (Walton, J., *Wireless World*, December, 1967, pp. 581–588), to whom I am indebted for the data on which Figure 10.12 is based. Below about 1kHz, the electrical output from the pick-up cartridge is limited by the allowable 0.01cm spacing between the grooves, and the resulting 0.005cm maximum permitted cutter excursion. Above a turnover frequency of around 1kHz the possible closeness of succeeding groove excursions, which control the output voltage, is determined by the shape of the rear face of the cutting head and the linear velocity of the stylus in the groove (which will depend on the track diameter). Beyond about 3kHz there is a limit imposed by the maximum groove accelerations which the cutter head can generate or the pick-up stylus can follow. These are also limited by the groove velocity, and thus depend on the track diameter.

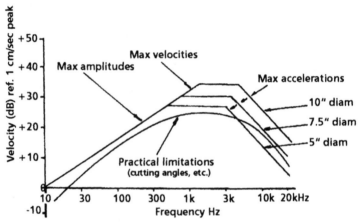

Figure 10.12
Maximum recorded levels

Walton considered that, under optimum conditions, and at a 7.5″ groove diameter, a groove modulation velocity equivalent to +30dB at 2kHz, with respect to 1cm/s, was the maximum which would be cut. The Shure Corporation, which made a speciality of the manufacture of pick-up cartridges with very high record groove tracking ability, claimed that its best models were capable of following a groove modulation of +40dB ref. 1 cm/s at 2kHz, decreasing with increasing frequency. They obviously considered that the musical bells track on their stereo test record, which was cut at a recording level of +25dB at 10kHz, would present a formidable challenge to their competitors – a consideration which lends support to Walton's curve of practical output limits in Figure 10.12.

From this analysis, if a recorded velocity of 5cm/s is taken as the normal maximum mid-band signal, the limitations of the recording process will ensure that the maximum modulation velocity which the pick-up will encounter on a very heavily modulated part of the recording will not exceed some 30cm/s. This practical limitation on the pick-up cartridge output voltage was also shown by Wolfenden (Wolfenden, B.S., *Wireless World*, December 1976, p. 54) using data from Shure, and by Kelly (Kelly, S., *Wireless World*, December 1969, pp. 548–555) who had confirmed, by experiment, the magnitude of any feasible overload margin.

However, with the same kind of logic which persuades ordinary, sensible motorists that a car which is capable of 150mph is a much better proposition – in a country where the speed limit is 70mph – than one which will only do 90mph, many circuit designers consider it necessary to provide RIAA stage voltage overload margins of 20–30˜, or greater. Unfortunately, by praising such achievements, reviewers perpetuate the belief that this style of design is both good and necessary. Clearly, some type of gain switching is necessary if the preamp is to cope with the voltage output levels from both MM and MC type pick-up cartridges, but these cartridges will require switch selection of their load resistance anyway, between, say, 47k and 100R, so the provision of a switchable choice between low impedance/high gain, and high impedance/low gain options has become normal practice. In reality, the highest output voltage peaks are likely to be those associated with dust particles, scratches and other blemishes on the surface of the disc, and while it is obviously desirable that such

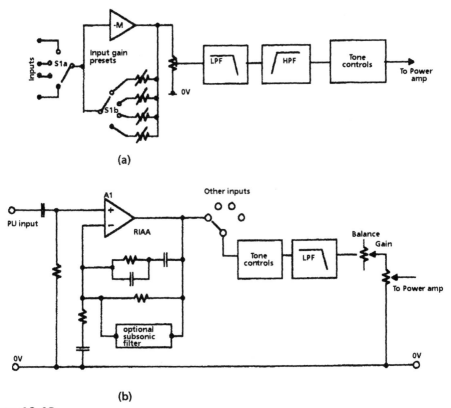

(a)

(b)

Figure 10.13
Basic preamp layouts

sudden input voltage excursions do not lead to a prolonged paralysis of the amplifier, clipping which is very brief in duration may be acceptable to the listener.

A typical op. amp. gain stage, fed from a ±15V supply, is capable of a 9.5V rms undistorted output voltage swing. It should therefore be capable of handling, without clipping, the signal from any disc replay amplifier whose normal maximum output level is less than 1.5V rms. For the greatest freedom from signal overload, the gain control should be placed as far forward in the signal chain as possible, as shown in schematic form in Figure 10.13a. Unfortunately, with this layout the total noise contributions from the following stages will be present all the time, regardless of gain control settings. The alternative option, shown in Figure 10.13b, is to place the RIAA equalisation stage between the pick-up input and the input selector switching. Any tone controls or filters, if used, can then be based on unity gain blocks, which if they use the same type of gain stage (such as a low noise op amp) as that used in the RIAA module itself, will not overload if the RIAA module does not. With this type of layout, the preamplifier noise contribution will decrease to zero as the signal level is reduced.

Tone Controls and Filters

In the earlier days of audio, it was taken for granted that there would be shortcomings both in the tonal quality and in the signal to noise ratio of all of the input signals. There would also be deficiencies in the frequency response characteristics of both the input transducers and the loudspeakers, and it was expected that any high quality preamplifier unit would offer comprehensive facilities to allow correction of the electrical output signal. However, there has been a continuous improvement in the quality of the input signal, whether obtained from compact disc, LP record, FM tuner or cassette tape. Similarly, there has been a continuing improvement in LS performance, so that, for example, the bass output is only limited by the physical dimensions of the LS cabinet, or of the listening room, and additional bass boost would be pointless. Faced with this situation, many amplifier manufacturers no longer offer anything other than flat frequency response amplifier units.

While the flat frequency response approach is probably adequate for most purposes, I feel that there are still a few remaining situations where, for example, high pass rumble filters are useful – such as when a new CD has been mastered from an archive source, and contains significant amounts of LF noise from mechanical sources, or where a blameless recording suffers from low frequency traffic noise intrusions into the notionally silent recording studio – or where the tonal balance chosen by the recording engineer differs from that preferred by the listener. In this case some form of frequency response tilt control would probably be useful. I have therefore shown two design examples of these. The first of these, due to Quad, is shown in Figure 10.14, and the second, due to Bingham (Bingham, J., *Hi-Fi News and Record Review*, December 1982, pp. 64–65) is shown in Figure 10.15. I have also shown the layouts for four steep cut filter circuits in Figure 10.16.

The filter circuits shown in Figures 10.16a and 10.16b use my own bootstrap design (Linsley Hood, J.L., *Electronic Engineering*, July 1976, pp. 55–58) and give a

−18dB/octave frequency response. That of Figure 10.16a is a 50Hz high pass layout, and that of 10.16b is a 10kHz low pass circuit. The filter circuits shown in Figures 10.16c and 10.16d have the same cut-off frequencies, but use the Sallen and Key layout (Sallen, R.P., and Key, E.L., *IRE Trans. Circuit Theory*, March 1955, pp. 40–42) which gives a −12dB/octave response. For both filters the turnover frequency is given by the formula:

$$f_T = 1/[2\pi\sqrt{(C1C2R1R2)}]$$

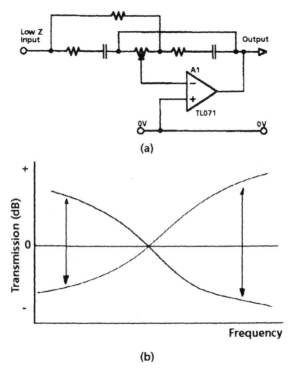

Figure 10.14
The 'Quad' slope control

The turnover frequencies can be changed, up or down, by proportional alterations to the resistor or capacitor values. For example, if the values of C1, C2 and C3, in Figure 10.16a, are increased by the factor 5/3, the turnover frequency of this filter will be reduced to 30Hz. Similarly, if the values of the resistors, R1, R2 and R3, are halved, the turnover frequency of Figure 10.16b will be increased to 20kHz, and so on.

Since the bulk of modern signal sources are not much in need of remedial treatment, I have not included technical descriptions of the wide variety of tone control arrangements which have been evolved over the years. For those seeking further information on these, I would recommend my book *Audio Electronics*, (Butterworth-Heinemann, 1965, Chapter 4, pp. 169–189), or an earlier article of mine in *Electronics World + Wireless World* (July 1990, pp. 636–639).

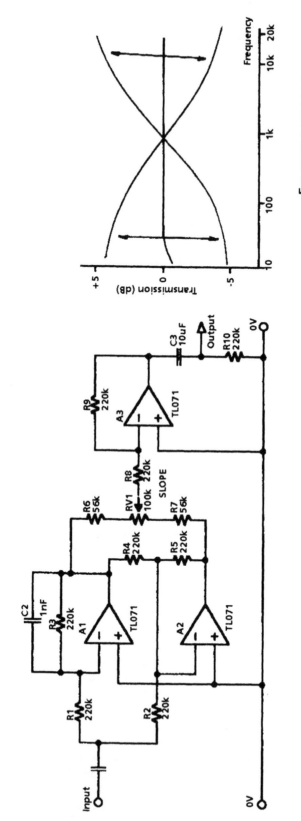

Figure 10.15
Bingham's tilt control

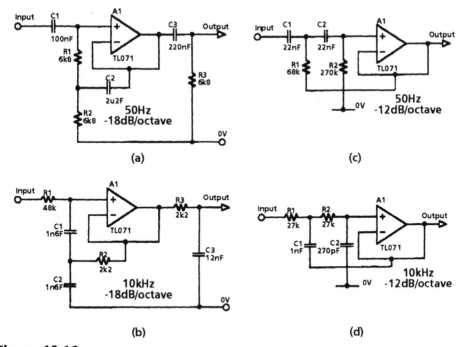

Figure 10.16
Steep cut filter circuits

Signal Channel Switching and Remote Controls

The choice between input channels can be made by purely mechanical means, such as a simple rotary switch or a suitable push-button sliding contact arrangement. The only normal difficulty with this system is that the knob or the push-buttons which control the channel selection will be on the front of the preamp box, while the signal inputs between which the electrical choice is to be made all go to sockets on the back face of the box, so that the internal connecting wires which join the one to the other must pass over other parts of the signal circuitry, and may need to be well screened. If the switching element is a relay (preferably sealed, and inert gas filled), or a suitable semiconductor switch, of the kind I have shown in Figure 10.17, the input switches can be located close to the input sockets. It will also allow channel selection to be made by an external voltage, and this will allow the convenience of remote control, perhaps by the use of an infra-red transmitter/receiver assembly of the type widely used for similar purposes with TV sets.

Most of the plain on/off types of signal switching, based on discrete components or CMOS analogue switches, of the type I have shown in Figure 10.17 offer acceptably low distortion characteristics up to, say, a 1V rms signal level. In the case of the series connected FET switch, shown in Figure 10.17c, the value of R_{in} should be high in relation to the 'on' resistance of the FET to swamp any signal induced changes in the channel resistance. In the case of the CMOS analogue switches shown in schematic form in Figure 10.17d, these devices normally are designed to operate with logic level input control voltages, and do not significantly distort input signals so long as their peak-to-peak values lie comfortably between the limits set by the DC supply rails. The extent of the distortion introduced by these devices depends on the output load (which should, ideally, be very high) and on the quality (and cost) of the component itself.

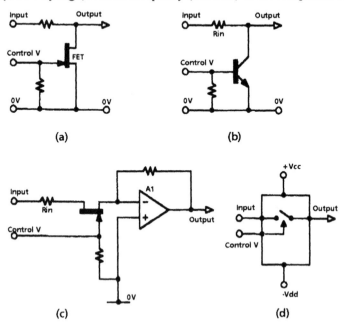

Figure 10.17
Voltage controlled switches

Voltage Controlled Gain Systems

While there are a number of low distortion switching circuits available, the techniques proposed for a continuous, voltage controlled, alteration of the amplifier gain are usually much less satisfactory. For simple, low cost signal-level control the FET arrangement of Figure 10.18a, in which the FET and the input resistance, R1, form a simple resistive attenuator, is fairly widely used. As it stands, this would have a THD figure, at 1V rms and 1kHz output, of some 10–15%. In the Siliconix file note, AN73–1, it is proposed that negative feedback should be derived from the drain circuit, by way of R2 and R3 (Siliconix suggest that the THD is lowest when R2 = R3 $10(R1//R_{ds}//R_L)$). This will reduce the THD to about 1–1.5% and although this figure decreases as the signal level is reduced, it is still very much higher than would be acceptable in any Hi-Fi context.

A commercially available alternative to the discrete component (e.g. FET) voltage controlled attenuator is marketed as the Motorola MC3340P IC, a device which was widely used in TV remote control systems. This uses two long-tailed pair circuits in which the gain is controlled by the tail current. By connecting the non-inverting and inverting outputs in antiphase the second harmonic distortion is partially cancelled. The claimed performance offers a 90dB possible attenuation ratio, a bandwidth of 1MHz, and a THD of 0.2%, worsening to 3% as the attenuation is increased.

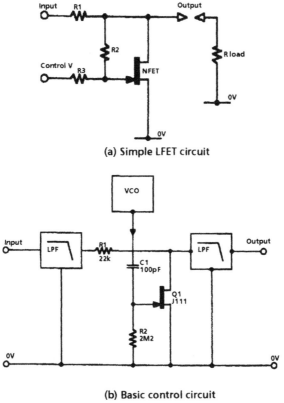

(a) Simple LFET circuit

(b) Basic control circuit

Figure 10.18
Attenuator circuits

I described the various other available techniques in an earlier article (Linsley Hood, J.L., *Electronics World + Wireless World*, April 1995, pp. 320–321), and suggested the use of a very low distortion discrete component design (intended for use in a dynamic range control circuit (Linsley Hood, J.L., *Electronics World and Wireless World*, November 1995, pp. 938–940). This gain control circuit was based on a switch-mode attenuator, shown in schematic form in Figure 10.18b. In this circuit a high frequency (180kHz) rectangular voltage waveform is used to control the proportion of the time in which the channel is open. This, in turn, is controlled by the 'on to off' ratio of the switching waveform, which can be changed, within the range 5% : 95% to 95% : 5% by a DC input voltage to the waveform generator circuit. This gives an attenuation ratio of about 30dB, and a THD – at 1V rms and 10Hz–10kHz – of less than 0.005%, mainly due to the steep-cut low pass filters used to remove signal components within 20kHz of the switching frequency.

In their application note AN–105, PMI propose an improved version of the system used by Motorola, of which I have shown the circuit in Figure 10.19. The THD introduced by this circuit depends on the cancellation of the distortion, mainly second harmonic, introduced by the controlled transistors, Q1–Q4. For this to be adequate, there must be an exceedingly close match between all four transistors, and PMI show this as an application for their MAT–04 monolithic transistor array. PMI claim a THD value of <0.03% at 3V rms output, for this circuit. Inevitably, such a useful discrete component layout will be produced in due course as an IC module, and this has been done by Analog Devices, Inc., in their SSM–2108T and SSM–2118T ICs, which have a claimed THD in the range 0.013–0.006% at 3V rms depending on operating conditions.

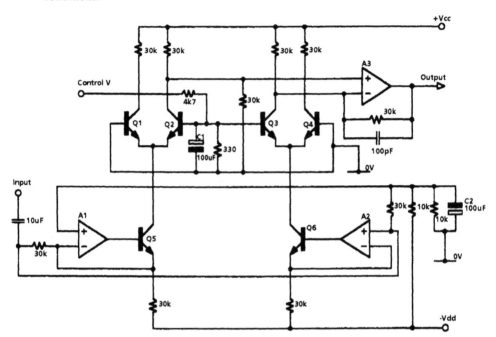

Figure 10.19
Voltage controlled gain stage, due to PMI

If a step-type gain adjustment is acceptable, a simple option is shown in the layout of Figure 10.20, in which the output from a point on a resistive divider chain is selected by one or other of a series of analogue switches. Since the sequential operation of these switches can be done by the use of a voltage controlled dot/bar driver IC, such as the LM3914, there would be little difficulty or expense in assembling the hardware for this type of relatively low distortion voltage operated volume control.

Digital Inputs

It has become customary for the manufacturers of compact disc players to provide a direct digital output from the disc replay, usually as the RF signal derived from the CD replay chain, shown, for reference, in Figure 10.21. This RF output will usually be taken at a point following the 14–8 decoder, but before the Cross-Interleave Reed-Solomon code (CIRC) error correction stage shown in the schematic replay layout of Figure 10.21. This allows the manufacturer of specialist digital to analogue signal-processing hardware to offer higher quality D/A conversion systems, more jitter-free clock regeneration circuitry (which could be very important in high precision D/A decoding), more competent error correction regimes and better anti-aliasing filtration than might be offered in a budget-priced CD player. It is argued that these latter signal handling stages probably contribute more to the overall tonal quality of the CD player than the mechanical platform itself or its control hardware.

In general, the RF output signal will be presented as a simple coaxial cable output, but sometimes in top of the range units, this signal may also be presented as an optical fibre output to avoid possible electrical interference or signal degradation in the link between the CD player and the A/D decoder.

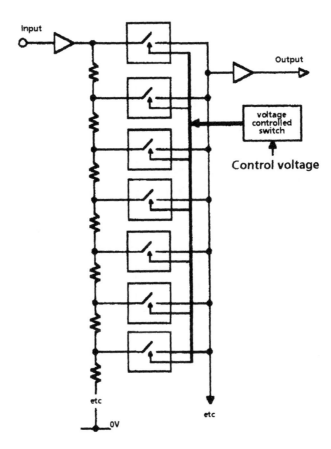

Figure 10.20
Step-type gain control

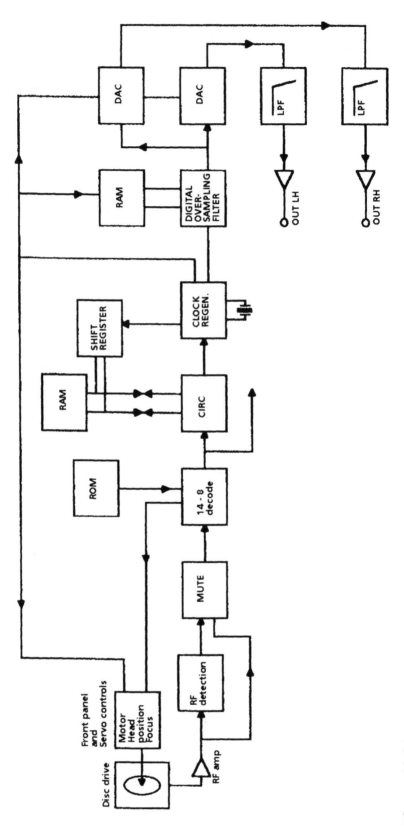

Figure 10.21
CD replay schematic layout

CHAPTER 11

POWER SUPPLIES

Active systems such as audio amplifiers operate by drawing current from some voltage source – ideally with a fixed and unvarying output – and transforming this into a variable voltage output which can be made to perform some useful function, such as driving a loudspeaker or some further active or passive circuit arrangement. For most active systems, the ideal supply voltage would be one having similar characteristics to a large lead-acid battery: a constant voltage, zero voltage ripple and a virtually unlimited ability to supply current on demand. In reality, considerations of weight, bulk and cost would rule out any such Utopian solution and the power supply arrangements will be chosen, with cost in mind, to match the requirements of the system they are intended to feed. However, the characteristics of the power supply used with an audio amplifier have a considerable influence on the performance of the amplifier, so this aspect of the system is one which cannot be ignored.

High Power Systems

In the early days of valve operated audio systems, virtually all of the mains powered DC power supply arrangements were of the form shown in Figure 11.1a and the only real choice open to the designers was whether they used a directly heated rectifier, such as a 5U4 or an indirectly heated one such as a 5V4 or a 5Z4. The indirectly heated valve offered the practical advantage that the cathode of the rectifier would heat up at roughly the same rate as that of the other valves in the amplifier, so there would not be an immediate switch-on no-load voltage surge of 1.4" the normal HT supply output voltage. With a directly heated rectifier this voltage surge would always appear in the interval between the rectifier reaching its operating temperature, which might take only a few seconds, and the thirty seconds or so which the rest of the valves in the system would need to come into operation and start drawing current. Using an indirectly heated rectifier would avoid this voltage surge and would allow lower working voltage components to be used with safety in the rest of the amplifier. This would save cost. On the other hand, the directly heated rectifier would have a more efficient cathode system, and would have a longer working life expectancy.

Although there are several other reasons for this – such as the greater ease of manufacture, by the use of modern techniques, of large value electrolytic capacitors, or the contemporary requirement that there shall be no audible mains hum in the amplifier output signal due to supply line AC ripple – it is apparent that the capacitance values used in the smoothing, decoupling and reservoir capacitors in traditional valve amplifier circuits are much smaller than in contemporary systems which operate at a lower output voltage. The main reason for this is that the stored energy in a capacitor is defined by the relationship:

$$E_c = \tfrac{1}{2}CV^2$$

where E_c is the stored energy, in joules, C is the capacitance, in farads, and V is the applied voltage. This means that there is as much energy stored in an 8μF capacitor, charged to 450V, as there is in a 400μF capacitor charged only to 64V. Since the effectiveness of a decoupling capacitor in avoiding the transmission of supply line rubbish, or a power supply reservoir capacitor in limiting the amount of ripple present on the output of a simple transformer/rectifier type of power supply, depends on the stored charge in the capacitor, its effectiveness is very dependent on the applied voltage – as is the discomfort of the electrical shock which the user would experience if he inadvertently discharged such a charged capacitor through his body.

Solid State Rectifiers

The advent of solid state rectifiers – nowadays almost exclusively based on silicon bipolar junction technology – effectively caused the demise of valve rectifier systems, although for a short period, prior to the general adoption of semiconductor rectifiers, gas-filled rectifiers, such as the 0Z4, had been used, principally in car radios, in the interests of greater circuit convenience because, in these valves, the cathode was heated by reverse ionic bombardment, so no separate rectifier heater supply was required. The difficulties caused by the use of these gas-filled rectifiers were that they had a relatively short working life and that they generated a lot of RF noise. This RF noise arose because of the very abrupt transition of the gas in the cathode/anode gap of the rectifier from a non-conducting to a conducting state. The very short duration high current spikes this caused shock-excited the secondary windings of the transformer – and all its associated wiring interconnections – into bursts of RF oscillation, which caused a persistent 100–120Hz rasping buzz, called modulation hum, to appear in the audio output.

The solution to this particular problem was the connection of a pair of capacitors, shown as C1 and C2 in Figure 11.1a, across the transformer secondary windings to re-tune any shock-excited RF oscillation into a lower, and less invasive, frequency band. Sometimes these modulation hum prevention capacitors are placed across the rectifiers or across the mains transformer primary winding, but they are less effective in these positions. With modern, low conduction resistance, semiconductor diodes, low equivalent series resistance (ESR) reservoir capacitors and low winding resistance (e.g. toroidal) transformers, this problem can still arise, and the inclusion of these capacitors is a worthwhile and inexpensive precaution. The circuit layout shown in Figure 11.1b is the PSU arrangement used in most contemporary valve amplifiers. For lower voltages a wider range of circuit layouts are commonly used, also shown in Figure 11.1.

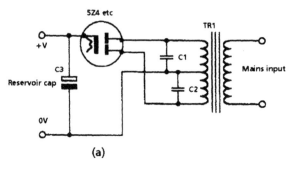

(a)

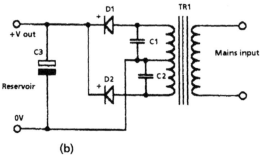

(b)

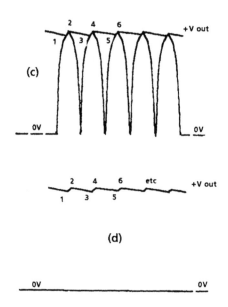

(c)

(d)

Output waveform with f-w recification

Figure 11.1
Full-wave rectifier systems

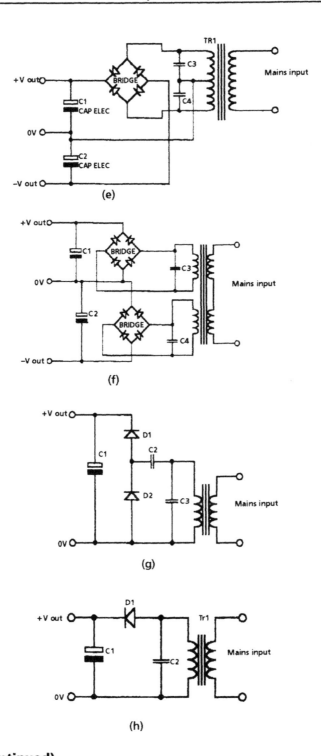

Figure 11.1 (continued)
Full-wave rectifier systems

Music Power

In their first flush of enthusiasm for solid state audio amplifiers, their manufacturers and their advertising copy writers collectively made the happy discovery that most inexpensive audio amplifiers powered by simple supply circuits – such as that shown in Figure 11.1b – would give a higher power output for short bursts of output signal, such as might, quite reasonably, be expected to arise in the reproduction of music, than they could give on a continuous sine-wave output. This short-duration, higher output power capability was therefore termed the music power rating, and, if based on a test in which, perhaps, only one channel was driven for a period of 100 milliseconds every second, would allow a music power rating to be claimed which was double that of the power which would be given on a continuous tone test, in which both channels are driven simultaneously (the so-called rms output power rating).

Influence of Signal Type on Power Supply Design

Although this particular method of specification enhancement is no longer widely used, its echoes linger on in relation to modern expectations for the performance of Hi-Fi equipment. The reason for this is that, in the earlier years of recorded music reproduction there were no such things as pop groups, and most of those interested in improving the quality of recording and replay systems were people such as Peter Walker of Quad or Gerald Briggs of Wharfedale loudspeakers, whose spare-time musical activities were as an orchestral flautist and a concert pianist, and whose interests, understandably, were almost exclusively concerned with the reproduction, as accurately as possible, of classical music. Consequently, when improvements in reproduction were attempted, they were in ways which helped to enhance the perceived fidelity in the reproduction of classical music, and the accuracy in the rendition of the tone of orchestral instruments. In general, this was easier to achieve if the electronic circuitry was fed from one or more accurately stabilised power supply sources, though this would nearly always mean that such power supplies would have, for reasons of circuit protection, a fixed maximum current output. While this would mean that the peak power and the rms power ratings would be the same, it also meant that there would be no reserve of power for sudden high level signal demands – a penalty which the tonal purists were prepared to accept as a simple fact of life.

However, times change, and Hi-Fi equipment has become more easy to accommodate, less expensive in relative terms, and much more widely available. Also, there has been a considerable growth in the purchasing power of those within relatively youthful age bracket, most of whose musical interests lie in the various forms of pop music – preferably performed at high signal levels – and it is for this large and relatively affluent group that most of the Hi-Fi magazines seek to cater. (If an example of the balance of contemporary musical taste is needed, consider the musical programme chosen by Martin Colloms and *Hi-Fi News* for the amplifier comparison trial referred to in Chapter 9, in which ten of the sixteen test pieces were drawn from the pop idiom.)

The ways in which these popular musical preferences influence the design of audio amplifiers and their power supplies relate, in large measure, to the peak short-term output current which is available, since one of the major instruments in any pop ensemble will be a string bass, whose sonic impact and attack will depend on the ability of the amplifier and power supply to drive large amounts of current into the LS load, and it must do this without causing any significant increase in the ripple on the DC supply lines or any loss of amplifier performance due to this cause.

A further important feature for the average listener to a typical pop ensemble is the performance of the lead vocalist, commonly a woman, the clarity of whose lyric must not be impaired by the high background signal level generated by the rest of the group. Indeed, with much pop music, with electronically enhanced instruments, the sound of the vocalist – though also electronically enhanced – is the nearest the listener will get to a recognisable reference sound. This clarity of the vocal line demands both low intermodulation distortion levels and a complete absence of peak-level clipping.

The designer of an amplifier which is intended to appeal to the pop music market must therefore ensure that the equipment can provide very large short duration bursts of power, that the power supply line ripple level, at high output powers, must not cause problems to the amplifier, and that, when the amplifier is driven into overload, it copes gracefully with this condition. The use of large amounts of NFB, which causes hard clipping on overload, is thought to be undesirable. Similarly, the effects of electronic (i.e. fast acting) output transistor current limiting circuitry (used very widely in earlier transistor audio amplifiers) would be quite unacceptable for most pop music applications, so alternative approaches, mainly based on more robust output transistors, must be used instead.

In view of the normal lack in much pop music of any identifiable reference sound source – such as would be provided by the orchestral or acoustic keyboard instruments in classical music forms – a variety of descriptive terms has emerged, to indicate the success or otherwise of the amplifier system in providing attractive reproduction of the music. Terms such as exciting or giving precise image location or vivid presence or having full sound staging or blurred or transparent are colourful and widely used in performance reviews, but they do not help the engineer in his attempts to approach more closely to an ideal system performance – attempts which must rely on engineering intuition and trial and error.

High Current Power Supply Systems

In order for the power supply system to be able to provide high output currents for short periods of time, the reservoir capacitor, C3 in Figure 11.1b, must be large, and have a low ESR (equivalent series resistance) value. Ideally, the rectifier diodes used in the power supplies should have a low conducting resistance, the mains transformer should have low resistance windings and low leakage inductance, and all the associated wiring – including any PCB tracks – should have the lowest practicable path resistance. The output current drawn from the transformer secondary winding, to replace the charge lost from the reservoir capacitor during the previous half cycle of discharge, occurs in brief, high current bursts, in the intervals between the points on the input voltage waveform labelled 1 and 2, 3 and 4, 5 and 6 and so on, shown in Figure 11.1c. This leads to an output ripple pattern of the kind shown in Figure 11.1d.

Unfortunately, all of the measures which the designer can adopt to increase the peak DC output current capability of the power supply unit will reduce the interval of time during which the reservoir capacitor is able to recharge. This will increase the peak rectifier/reservoir capacitor recharge current and will shorten the duration of these high current pulses. This increases the transformer core losses, and both the transformer winding and lamination noise, and also increases the stray magnetic field radiated from the transformer windings. All of these factors increase the mains hum background, both electrical and acoustic, of the power supply unless steps are taken – in respect of the physical layout, and the placing of interconnections – to minimise it. The main action which can be taken is to provide a very large mains transformer, apparently excessively generously rated in relation to the output power it has to supply, in order that it can cope with the very high peak secondary current demand without mechanical hum or excessive electromagnetic radiation. Needless to say, the mains transformer should be mounted as far away as possible from regions of low signal level circuitry, and its orientation should be chosen so that its stray magnetic field will be at right angles to the plane of the amplifier PCB.

Half-wave and Full-wave Rectification

Because the reservoir capacitor recharge current must replace the current drawn from it during the non-conducting portion of the input cycle, both the peak recharge current and the residual ripple will be twice as large if half-wave rectification is employed, such as that shown in the circuit of Figure 11.1h, in which the rectifier diode only conducts during every other half cycle of the secondary output voltage, rather than on both cycles as would be the case in Figure 11.1b. A drawback with the layouts of both Figure 11.1a and 11.1b is that the transformer secondary windings only deliver power to the load every other half cycle, which means that when they do conduct, they must pass twice the current they would have had to supply in, for example, the bridge rectifier circuit shown in Figure 11.1e. The importance of this is that the winding losses are related to the square of the output current ($P = i^2R$) so that the transformer copper losses would be four times as great in the circuit of Figure 11.1b as they would be for either of the bridge rectifier circuits of Figure 11.1f. On the other hand, in the layout of Figure 11.1b, during the conduction cycle in which the reservoir capacitor is recharged, only one conducting diode is in the current path, as compared with two, in the bridge rectifier set-ups.

Many contemporary audio amplifier systems require symmetrical +ve and –ve power supply rails. If a mains transformer with a centre-tapped secondary winding is available, such a pair of split-rail supplies can be provided by the layout of Figure 11.1e, or, if component cost is of no importance, by the double bridge circuit of Figure 11.1f. The half-wave voltage doubler circuit shown in Figure 11.1g is used mainly in low current applications where its output voltage characteristic is of value – such as perhaps a higher voltage, low-current source for a three-terminal voltage regulator.

DC Supply Line Ripple Rejection

Avoidance of the intrusion of AC ripple or other unwanted signal components from the DC supply rails can be helped in two ways – by the use of voltage regulator circuitry to maintain these rails at a constant voltage, or by choosing the design of the amplifier circuitry which is used so that there is a measure of inherent supply line signal rejection. In a typical audio power amplifier, of which the voltage amplifier stage is shown in Figure 9.11, there will be very little signal intrusion from the +ve supply line through the constant current source, Q6 and Q7, because this has a very high output impedance in comparison with the emitter impedance of Q1 and Q2, so any AC ripple passing down this path would be very highly attenuated. On the other hand, there would be no attenuation of rubbish entering the signal line via R5, so that, in a real-life amplifier, R5 would invariably be replaced by another constant current source, such as that arranged around Q7 and Q8.

For the negative supply rail, the cascode connection of Q10 would give this device an exceedingly high output impedance, so any signal entering via this path would be very heavily attenuated by the inevitable load impedance of the amplifier. Similarly, the output impedance of the cascode connected transistors Q3 and Q4 would be so high that the voltage developed across the current mirror (Q5 and Q6) would be virtually independent of any –ve rail ripple voltage. In general, the techniques employed to avoid supply line intrusion are to use circuits with high output impedances wherever a connection must be made to the supply line rails. In order of effectiveness, these would be a cascode connected FET or bipolar device, a constant current source, a current mirror or a decoupled output – such as a bootstrapped load. HT line decoupling, by means of an LF choke or a resistor and a shunt connected capacitor, such as R2 and C2 in Figure 4.2, was widely used in valve amplifier circuitry, mainly because there were few other options available to the designer. Such an arrangement is still a useful possibility if the current flow is low enough for the value of R2 to be high, and if the supply voltage is high enough for the voltage drop across this component to be unimportant. It still suffers from the snag that its effectiveness decreases at low frequencies where the shunt impedance of C2 begins to increase.

Voltage Regulator Systems

Electronic voltage regulator systems can operate in two distinct modes, each with their own advantages and drawbacks: shunt and series. The shunt systems operate by drawing current from the supply at a level which is calculated to be somewhat greater than maximum value which will be consumed by the load. A typical shunt regulator circuit is that shown in Figure 11.2a, in which the regulator device is an avalanche or Zener diode, or, for low current, high stability requirements a two-terminal band-gap element. Such simple circuits are normally only used for relatively low current applications, though high power avalanche diodes are available. If high power shunt regulators are needed a better approach is to use a combination of avalanche diode and power transistor, as shown in Figure 11.2b. The obvious snag is that in order for such a system to work, there must be a continuous current drain which is rather larger than the maximum likely to be drawn by the load, and this is wasteful. The main advantages of the shunt regulator system are that it is simple, and that it can be used even when the available supply voltage is only a little greater than the required output voltage. Avalanche and Zener diodes are noisy, electrically speaking, though their noise can be lessened by connecting a low ESR capacitor in parallel with them. For applications where only a low voltage is needed, its actual value is not very important but a low circuit noise is essential, a simple arrangement is to use a string of silicon diodes, as shown in Figure 11.2c. Each of these diodes will have a forward direction voltage drop of about 0.6V, depending on the current flowing though them. Light emitting diodes have also been recommended in this application, for which a typical forward voltage drop would be about 2.4V, depending on the LED type and its forward current. All of these simple shunt regulator circuits will perform better if the input resistor (R1) is replaced by a constant current source, shown as CC1.

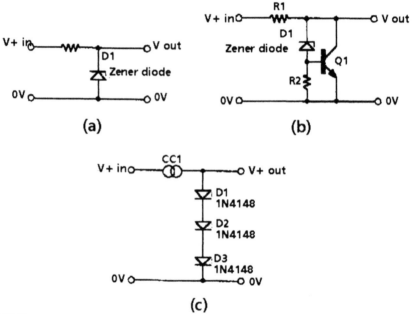

(a)

(b)

(c)

Figure 11.2
Simple shunt regulators

Series Regulator Layouts

The problem with the shunt regulator arrangement is that the circuit must draw a current which is always greater than would have been drawn by the load on its own. This is an acceptable situation if the total current levels are small, but this would not be tolerable if high output power levels were involved. In this situation it is necessary to use a series regulator arrangement, of which I have shown some simple circuit layouts in Figure 11.3. The circuit of Figure 11.3a forms the basis for almost all this type of regulator circuit, with various degrees of elaboration. Essentially, it is a fixed voltage source to which an emitter–follower has been connected to provide an output voltage (that of the Zener diode less the forward emitter bias of Q1) at a low output impedance. The main problem is that, for the circuit to work, the input voltage must exceed the output voltage – the difference is termed the drop-out voltage – by enough voltage for the current flow through R1 to provide both the necessary base current for Q1 and also enough current through D1 for D1 to reach its reference voltage. Practical considerations require that R1 shall not be too small. In a well–designed regulator of this kind, such as the 78xx series voltage regulator IC, the drop-out voltage will be about 2V.

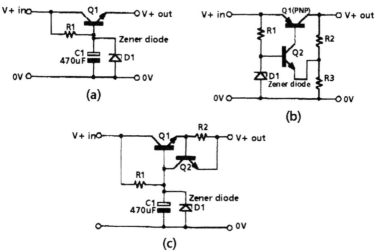

Figure 11.3
Simple series regulators

This drop-out voltage can be reduced by reversing the polarity of Q1, as shown in Figure 11.3b, so that the required base input current for Q1 is drawn from the 0V rail. This arrangement works quite well, except that the power supply output impedance is much higher than that of Figure 11.3a, unless there is considerable gain in the NFB control loop. In this particular instance Q2 will conduct, and feed current into Q1 base until the voltage developed across R3 approaches the voltage on the base of Q2, when both Q2 and Q1 will be turned off. By augmenting Q2 with an op amp, as I have shown in Figure 11.4, a very high performance can be obtained from this inverted type of regulator layout.

Over-current Protection

A fundamental problem with any kind of solid state voltage regulator layout, such as that of Figure 11.3a, is that if the output is short-circuited, the only limit to the current which can flow is the capacity of the input power supply, which could well be high enough to destroy the pass transistor (Q1). For such a circuit to be usable in the real world, where HT rail short-circuits can, and will, occur, some sort of over-current protection must be provided. In the case of Figure 11.3c, this is done by putting a resistor (R2) in series with the regulator output, and then arranging a further transistor (Q2) to monitor the voltage across this. If the output current demand is enough to develop a voltage greater than about 0.65V across R2, Q2 will conduct, and will progressively steal the base current from Q1.

In the inverted stabiliser circuit shown in Figure 11.4, R1 monitors the output current, and if this is large enough to cause Q1 to conduct, then the output voltage will progressively collapse, causing the PSU to behave as a constant current source, at whatever output voltage causes the load to draw the current determined by R1. (I know this protection technique works because this is the circuit I designed for my workshop bench power supply twenty years ago (Linsley Hood, J.L., *Wireless World*, January 1975, pp. 43–45), and it has been in use every working day since then, having endured countless inadvertent output short-circuits during normal use, as well as surviving my son having left it on overnight, at maximum current output, connected to a nickel-plating bath which he had hooked up, but which had inadvertently become short-circuited.) In the particular layout shown, the characteristics of the pass transistors used (Q3 and its opposite number) are such that no current/voltage combinations which can be applied will cause Q3 to exceed its safe operating area boundaries, but this is an aspect which must be borne in mind. Although I use this supply for the initial testing of nearly all my amplifier designs, it would not have an acceptable performance, for reasons given above, as the power supply for the output stage of a modern Hi-Fi amplifier.

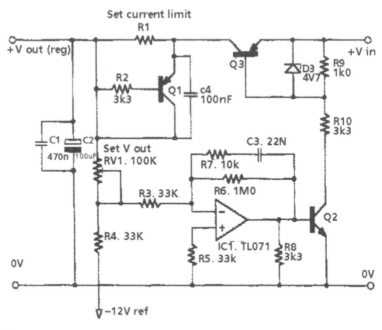

Figure 11.4
Series stabilised PSU

However, there is no such demand for a completely unlimited supply current for the voltage amplifier stages or the preamplifier supply rails, and in these positions, a high quality regulator circuit can be of considerable value in avoiding potential problems due to hum and distortion components breaking through from the PSU rails. Indeed, there is a trend, in modern amplifier design, to divide the power supplies to the amplifier into several separate groupings, one pair for the gain stages, a second pair for the output driver transistors, and a final pair of unregulated supplies to drive the output transistors themselves. Only this last pair of supplies need normally to be fed directly from a simple high current rectifier/reservoir capacitor type of DC supply system.

A further possibility which arises from the availability of more than one power supply to the power amplifier is that it allows the designer, by the choice of the individual supply voltages which are provided, to determine whereabouts in the power amplifier the circuit will overload, when driven too hard, since, in general, it is better if it is not the output stage which clips. This was an option of which I took advantage in my 80 watt power MOSFET design of 1984 (Linsley Hood, J.L., *Electronics Today International*, June 1984, pp. 24–31).

Integrated Circuit (Three Terminal) Voltage Regulator ICs

For output voltages up to ±24V, and currents up to 5 amperes, depending on voltage rating, a range of highly developed IC voltage regulator packages are now offered, having over-current (s/c), and thermal overload protection, coupled with a very high degree of output voltage stability, coupled with a typical >60dB input/output line ripple rejection. They are most readily available in +5V and +15V/–15V output voltages because of the requirements of 5V logic ICs and of IC op amps, widely used in preamplifier circuits, for which ±15V supply rails are almost invariably specified. Indeed, the superlative performance of contemporary IC op amps designed for use in audio applications is such an attractive feature that most audio power amplifiers are now designed so that the maximum signal voltage which is required from the pre amp is within the typical 9.5V rms output voltage available from such IC op amps.

Higher-voltage regulator ICs, such as the LM337T and the LM317T, with output voltages up to –37V and +37V respectively, and output currents up to 1.5A, are available, but where audio amplifier designs require higher voltage stabilised supply rails, the most common approaches are either to extend the voltage and current capabilities of the standard IC regulator by adding on suitable discrete component circuitry, as shown in Figure 11.5, or by assembling a complete discrete component regulator of the kind shown in Figure 11.6.

In the circuit arrangement shown for a single channel in Figure 11.5, a small-power transistor, Q1, is used to reduce the 55–60V output from the unregulated PSU to a level which is within the permitted input voltage range for the 7815 voltage regulator IC (IC2). This is one of a pair providing a ±15V DC supply for a preamplifier. A similar 15V regulator IC (IC1) has its input voltage reduced to the same level by the emitter-follower Q4, and is used to drive a resistive load (R7), via the control transistor, Q5. If the output voltage, and consequently the voltage at Q5 base, is too low, Q5 will conduct, current will be drawn from the regulator IC (IC1), and, via Q4, from the base of the pass transistor, Q2. This will increase the current through Q2 into the output load, and will increase the output voltage. If, however, the output voltage tends to rise to a higher level than that set by RV1, Q5 will tend towards cut-off, and the current drawn from Q2 base will be reduced, to restore the target output voltage level.

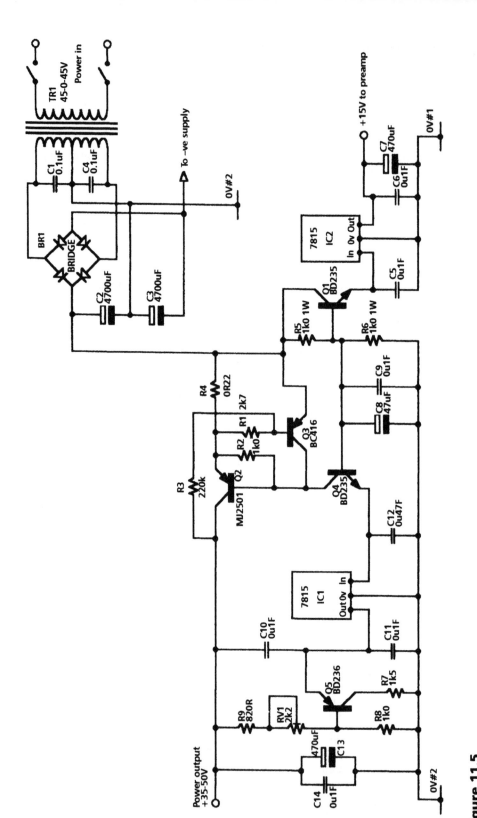

Figure 11.5
Stabilised PSU (one half only shown)

Over-current protection is provided by the transistor Q3 which monitors the voltage developed across R4, and restricts the drive to Q2 if the output current is too high. Safe operating area conformity is ensured by the resistor R3, which monitors the voltage across the pass transistor, and cuts off Q2 base current if this voltage becomes too high.

In the circuit of Figure 11.6, which is used as the power supply for an 80–100W power MOSFET audio amplifier – again only one channel is shown – a P-channel power MOSFET is used as the pass transistor, and a circuit design based on discrete components is used to control the output voltage. In this, transistor Q21 is used to monitor the potential developed across R33 through the R35/RV3 resistor chain. If this is below the target value, current is drawn through Q19 and R29, to increase the current flow through the pass transistor (Q17). If either the output current or the voltage across Q17 is too high, Q7 is cut off and there is no current flow through Q18 into Q17 gate.

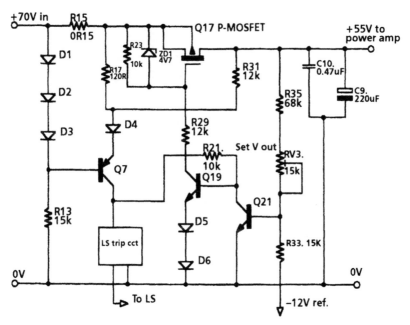

Figure 11.6
S/c protected PSU

This regulator circuit allows electronic shut-down of the power supply if an abnormal output voltage is detected across the LS terminals (due, perhaps, to a component failure). This monitoring circuit (one for each channel) is shown in Figure 11.7. This uses a pair of small-signal transistors, Q1 and Q2, in a thyristor configuration which, if Q2 is turned on, will connect Q1 base to the 0V rail, which, in turn, causes current to be drawn from Q2 base, which causes Q2 to remain in conduction even if the original input voltage is removed. The trip voltage will arise if an excess DC signal (e.g. >10V) appears across the LS output for a sufficient length of time for Q1 to charge to +5V. Returning to Figure 11.6, when the circuit trips, the forward bias voltage present on Q19 base is removed, and Q17 is cut off, and remains cut off until the trip circuit is reset by shorting Q2 base to the 0V rail. If the fault persists, the supply will cut out again as soon as the reset button is released. An electronic cut-out system like this avoids the need for relay contacts or fuses in the amplifier output lines. Relays can be satisfactory if they are sealed, inert gas-filled types, but fuseholders are, inevitably, crude, low cost components, of poor construction quality and with a variable and uncertain contact resistance. These are best eliminated from any signal line.

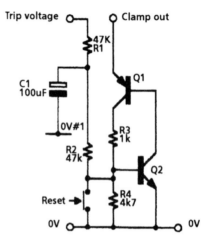

Figure 11.7
Trip circuit

Typical Contemporary Commercial Practice

The power supply circuit used in the Rotel RHB10 330 watt power amplifier is shown in Figure 11.8 as an example of typical modern commercial practice. In this design, two separate mains power transformers are used, one for each channel (the drawing only shows the LH channel – the RH one is identical) and two separate bridge rectifiers are used to provide separate ±70V DC outputs for the power output transistors and the driver transistors. This eliminates the distortion which might otherwise arise because of breakthrough of signal components from the output transistor supply rail into the low power signal channel. Similarly, the use of a separate supply system for each channel eliminates any power supply line induced L–R cross-talk which might impair stereo image positioning.

Battery Supplies

An interesting new development is the use of internally mounted rechargeable batteries as the power supply source for sensitive parts of the amplifier circuitry – such as low input signal level gain stages. Provided that the unit is connected to a mains power line, these batteries will be recharged during the time the equipment is switched off, but will be disconnected automatically from the charger source as soon as the amplifier is switched on.

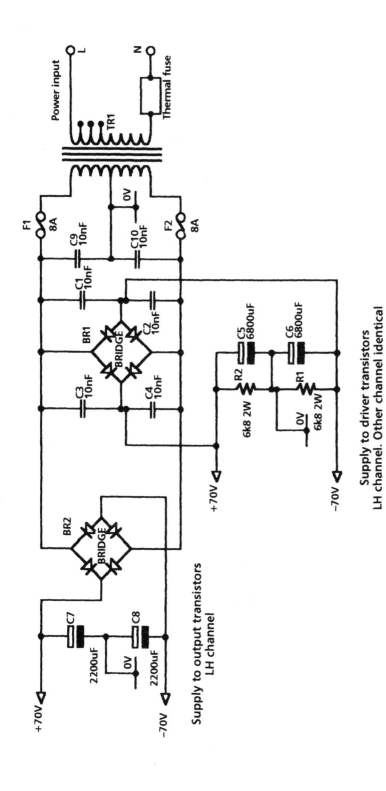

Figure 11.8
Rotel rhb10 PSU (only one channel shown)

Switch-mode Power Supplies

These are widely used in computer power supply systems, and offer a compact, high efficiency regulated voltage power source. They are not used in Hi-Fi systems because they generate an unacceptable level of HF switching noise, due to the circuit operation. They would also fail the requirement to provide high peak output current levels.

INDEX

D

E

Printed and bound by CPI Group (UK) Ltd, Croydon, CR0 4YY

03/10/2024

01040432-0015